AF352283

Information and Communication Technologies in the Context of Globalization

Information and Communication Technologies in the Context of Globalization

Evidence from Developing Countries

Kaushalesh Lal

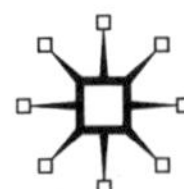

First published 2007 by
PALGRAVE MACMILLAN
Houndmills, Basingstoke, Hampshire RG21 6XS and
175 Fifth Avenue, New York, N.Y. 10010
Companies and representatives throughout the world

PALGRAVE MACMILLAN is the global academic imprint of the Palgrave Macmillan division of St. Martin's Press, LLC and of Palgrave Macmillan Ltd. Macmillan® is a registered trademark in the United States, United Kingdom and other countries. Palgrave is a registered trademark in the European Union and other countries.

ISBN 13: 978–0–230–53982–2   hardback
ISBN 10: 0–230–53982–3   hardback

A catalogue record for this book is available from the British Library.

A catalogue record for this book is available from the Library of Congress.

10   9   8   7   6   5   4   3   2   1
16   15   14   13   12   11   10   09   08   07

Printed and bound in Great Britain by
Antony Rowe Ltd, Chippenham and Eastbourne

Contents

Acknowledgements

The book is outcome of a project of UNU-MERIT entitled 'Globalization and the Adoption of Information and Communication Technologies in Developing Countries: A Cross-country Analysis'. Focus of the project has been on the adoption of ICTs in small and medium firms. Five countries, namely: Malaysia, India, Nigeria, Jamaica, and Costa Rica were covered in the study. Mr. Ong Seng Fook of the University Tunku Abdul Rahman coordinated Malaysian part of the study while Society for Economic and Social Research, New Delhi assisted in the survey of Indian sample firms. Prof. Femi Olokesusi of NISER was kind enough to supervise the field survey in Nigeria and provided insight of SMEs development in Nigeria. Prof. Ade Ogunrinade of University of Technology, Kingston contributed in a similar way with regard to Jamaica. Mr. Jose Pacheco Jimenez of Sanigest International, San Jose, Costa Rica and Mr. Federico Quesada Chaves of Instituto Latino Americano de Políticas Públicas, San Jose, Costa Rica coordinated Costa Rican component of the study. We would like to thank all the coordinators for their contribution in the project.

Greatest contribution to this study is attributed to the managing director of firms who spared their valuable time and shared information about their companies. We would like to specially thank Prof. Arup Mitra of Institute of Economic Growth, University of Delhi, India for excellent technical editing of the book.

I am grateful to Marc Vleugels for his administrative support throughout the project. Thanks are due to Ms. Eveline in de Braek and Ms. Maria Healy for their assistance in the project.

Kaushalesh Lal
May 2007

List of Figures

List of Tables

List of Appendix Tables

1
Globalization and SMEs: Introduction and Conclusion

1 Introduction

Industry is the engine of growth and one of the necessary conditions required to perform this role successfully include technological progress. For a long time several countries, particularly the developing ones, have struggled to determine appropriate technology and innovations, which promise high pay-offs. The advent of Information and Communication Technologies (ICTs) in this context has come as a technological revolution, which has enabled even some of the developing countries like India to make their presence felt in the world economy. It allows a reduction in co-ordination costs and leads to efficient electronic markets. Significant returns on ICT investments are found both in developing and developed countries. These benefits range from employment, to productivity gains, consumer surplus, and improvement in product quality.

Ample evidence exists to suggest substantive returns on IT systems, equipment and labour investments (Lichtenberg, 1995). Also, there is a strong relationship between IT and improvements in economic performance (Stiroh, 2001). Significant externalities emerge in the process of investment made in R&D due to complementary effects. In other words, as the level of 'general purpose technology' which is an outcome of adoption of ICTs, improves, it results in productivity growth across other sectors as well.

The other angle from which the contribution of ICTs to growth can be emphasized is infrastructure. Several studies have noted that infrastructure – social, financial and physical – contribute to total factor productivity growth across industries. This improves the level of technological efficiency as well the pace of technological progress (Mitra, 1999 and Mitra *et al.*, 2002). However, along with these components of infrastructure it is equally important to recognize the relevance of technological infrastructure, supplied by ICTs. Technological infrastructure, which may be broadly divided into three components, includes telecommunications, computing, and connectivity infrastructure (Oyelaran-Oyeyinka and Lal, 2006).

It may be noted that ICTs cut across types of activities. They can be applied in a wide variety of fields such as agriculture, health, education and training, manufacturing, services, transport, business, and environmental management (Lal, 2006). In the recent years in several developing countries much of the growth spur seems to have come from the services sector. And several economists have expressed concern in this regard as services growth is often viewed to be parasitic in nature. But a closer look at the composition of growth suggests that much of the services growth is taking place due to the growth of the dynamic components within services. And these dynamic components obviously include ICTs and tend to be the engine of growth. However, the developing countries cannot afford to ignore the importance of manufacturing or agriculture in the context of rapid economic growth. Hence, the application of ICTs in these sectors must be carried out aggressively which in return can contribute to productivity and faster economic growth. On the whole, industry is still the engine of growth and for industry to carry out its role successfully the ICTs have to undergo rapid expansion and progress.

2 Importance of SMEs

Within the industrial sector small and medium-sized enterprises (SMEs) are often regarded as the lifeblood of the economy in both developed and developing countries. The chapter on Malaysia argues that they help bring out innovation, increased productivity, competition and high value added activities. They are highly flexible and are able to explore and experiment new ideas and practices in the quest for efficiency and effectiveness. Lee (2003) argues that SMEs form an integral part of the value system as downstream suppliers or service providers to the large corporations in the overall production network. Realizing these crucial roles that SMEs play, governments are keen in designing appropriate industrial policies to further harness the potential of SMEs as a catalyst to economic growth of the countries.

Seow (1989) argues that SMEs contribute towards the national economy in five main areas. First, SMEs employ higher labour per capita investment. Although SMEs factories do not employ as much labour as the large manufacturing industries, in comparison to the capital invested in the business they employ more labour per dollar invested. So SMEs actually help resolve some of the labour problems. Second, SMEs assist tremendously towards increasing investment. The principle sources of finance for the SMEs are the owners own funds and those borrowed from other family members, friends and relatives and possibly some close business associates. Third, SMEs play a big role in generating entrepreneurship and creativity. The low cost of setting up a small manufacturing unit enables an enterprising worker not only to provide himself with a livelihood but also to offer employment to others. Therefore, SMEs serve as a training ground for developing the skills of industrial workers and entrepreneurs. These entrepreneurs who have been trained in basic skills usually are creative in their own ways and may branch into other related fields and thus SMEs produce creativity, and this creativity of the entrepreneurs can generate new products.

Fourth, SMEs are also a good training ground for skill development. The establishment of SMEs actually enables the small entrepreneurs to use their skill and knowledge, as well as to improve their technological capabilities. Therefore, this eventually accelerates the process of technology absorption and dispersion of economic benefits. And finally, SMEs provide back-up service for the large industries. This is because it may not be cost effective for the large industries to go for manufacturing component parts. It would be more economical if they have their component parts manufactured either by their subsidiaries or sub-contract it to various SMEs which are able to supply them with these components at a cheaper price and sometimes more efficiently.

With the advent of ICTs and improved transportation people can travel easily and communicate effectively and efficiently. The direct result of this is that people can work together without having to be physically together. Organizations find that they can easily outsource some of their work to smaller companies, thus encouraging the establishment of more service industries. Besides SMEs are flexible in responding to demand changes, which Beaver (2002) describes as enterprise culture.

3   Evolution of SMEs

The evolution of manufacturing sector has taken several years and it varies from country to country. As the chapter on Costa Rica brings out, the evolution can be analysed in different time periods. The first period, before 1960, is characterized by the predominance of the agricultural sector in the overall economy and the poor development of the manufacturing structure. During that period, two crops, coffee and banana, represented most of the exports. Then, during the second period, between 1960 and 1985, the Industrialization Model (IM) was actively implemented and consequently the manufacturing sector emerged as the engine of growth of the Costa Rican economy (6 per cent average growth rate experienced between 1960 and 1970). These first two periods comprise the pre-Structural Adjustment era. Finally, in the third stage, 1986 onwards, government adopted structural adjustment programme, trade liberalization, and competitiveness policies. In this period, although policy orientation mainly focused on the transformation of SMEs into externally competitive industries, share of the manufacturing sector in GDP practically remained constant.

Under the agricultural-based model, few products, particularly coffee and bananas, comprised the core of the national GDP and their participation in total exports was over 60 per cent. The dependence on two crops made the country highly vulnerable to fluctuations in prices and thus to external crisis in the balance of payments. Despite this fragility, practically no industrial policy was implemented during this period and the manufacturing sector was a series of small-dispersed firms with no intra and inter-sectoral links and poorly developed managerial capabilities.

Based on the ideas developed by the structuralist theory, spread in Latin America by the Economic Commission for Latin America and the Caribbean, the country adopted a new law, the well-known Act of Industrial Protection and Development (1959) that created a series of tax exemptions, subsidies, and tariffs to spur the development of the local industry. Later, with the incorporation of Costa Rica in the Central American Common Market, the adoption of the Central American Common Tariff provided the package of incentives by providing extra protection to the local infant industry and a bigger market for selling its products. The new legal and institutional framework provided positive results and thus the manufacturing sector became the most dynamic sector of the Costa Rican economy.

Data on the contribution of the manufacturing sector to national employment and wages also reflect a significant dynamism. Studies have confirmed that upward trend in employment was experienced by the manufacturing sector, although several concerns have been raised on the asymmetry between the dynamics of the manufacturing output and the pace of job creation in the manufacturing sector. Finally, in terms of remunerations, the sector contributed to 15 per cent of national remunerations.

Manufacturing sector exports and imports experienced a positive growth rate. Manufacturing exports almost quadrupled their participation in total exports while imports almost doubled their contribution. The Central American market became the most important destination by absorbing 70 per cent of the manufacturing exports. This market was so important that, if we take it off from the analysis, the results show that only 3 per cent of the small-scale production, 6 per cent of the medium-scale output and 8 per cent of the big firms' production was exported outside Central America.

Several structural problems and an obscure international context eroded the IM. So by the end of the 1970s the model experienced some problems. The 1981–82 crisis forced the adoption of a set of policy measures to stabilize the economy and to progressively implement a new development model. After 2 years of stabilization efforts, the country actively began the implementation of trade liberalization and pro-competitiveness policies as part of a broader package of structural adjustment measures.

The Structural Adjustment Program comprised a set of public policies aiming at reducing and eliminating those barriers that do not allow a full utilization of the economic resources. Part of the efforts was addressed to implementing a new trade model characterized by lower tariffs and export promotion. By lowering tariffs, the industrial sector would face an increasing competition from international firms, improving its level of competitiveness. By promoting exports, the sector would have enough incentives for diversifying its products. The application of fiscal subsidies and the creation of a wide range of trade-related institutions were the two most important tools used by the different governments to promote the shift from one commercial regime to the other. The results of this model in the evolution of the manufacturing sector are discussed in Chapter 4.

4   Constraint and supports for SMEs

There are a number of constraints, as brought out by the chapter on Jamaica, which limit the growth of SMEs. They include inadequate credit facilities and infrastructural and technological problems, hampering the development of the sector and limiting the growth, global competitiveness and the formation of linkages in the global commodity chain by SMEs.

Governments in several developing countries have taken initiatives to retain/strengthen competitiveness of SMEs in the era of globalization. These measures range from technological up-gradation support to new opportunities for human resource development, and marketing support. Technological support is not only limited to make SMEs aware of advanced technological development which has taken place in the world market but also to help them in acquisition of new technologies. One possible way to encourage firms to adopt new technology is through cost-reduction in accessing such technologies. This can be achieved by providing soft loans for importing new technologies. For export-oriented SMEs a reduction in custom duties might encourage the adoption of new technologies.

Support to SMEs in the era of globalization has become imperative due to entry of Multi-National Corporations (MNCs) in developing countries. It is well known that MNCs are equipped with several tangible and intangible assets. These assets could vary from efficient production technologies to brand name advantages. In order to compete even in the domestic market SMEs need to upgrade their business processes, which is virtually impossible for SMEs without active support from the government. This is primarily because neither SMEs have enough resources at their disposal to invest in new technologies nor do they have enough risk absorbing capacity.

Governments in developing countries are aware of these problems and many initiatives have been take to encounter the onslaught of globalization. However, only policy initiatives are not sufficient for success; there has to be willingness and eagerness to adopt new technologies in SMEs. The absence of such motivations might lead to misuse of policies and government resources as seen in several studies, which examined the effect of policy measures in India, Malaysia, and Jamaica on the performance of SMEs (Lal, 2004; Lee, 2003; McFarlane, 1997).

In Jamaica, for example, fiscal and credit support to SMEs are available in terms of lending institutions – some of which offer additional training and technology support. However, the extent of usage of these credit lines and technological support are not known since most SMEs remain outside the domain of the formal sector of the economy. The author evaluates the competitiveness of SMEs in Jamaica and their ability to use ICTs to enhance global competitiveness and add value in the global commodity chain.

5   New challenges for SMEs in the face of globalization

The process of globalization, as the chapter on Nigeria points out, is rapid, complex and full of uncertainties and expectations. It is obvious that only those countries that invest in science and technology (S&T) are able to integrate themselves better in the global system and to effectively respond to the challenges posed by globalization. Consequent upon this new era, the African continent in general and Nigeria in particular, are faced with new challenges in the process of socio-economic development. One dimension of this is how Nigeria can further integrate with the global economy through enhanced trade, and still cope with the prospect of unfettered economic growth and greater openness. This challenge is enormous especially given the current low level of penetration of processed and manufactured goods as well as services from Nigeria into the world trading system. Exports especially industrial goods from Nigeria and other developing nations are confronted with some competitive forces such as rivalry of competitors within its industry, threats of new entrants, threats of substitutes, the bargaining power of customers and the bargaining power of suppliers. The awesome capabilities of the various ICTs to confront some of these forces have been proved in a number of countries such as the United States, Brazil, Guatemala, and Tunisia (UNCTAD, 2003, 2004a).

The available data show that international trade in ICT goods and services has grown in recent years at a faster rate than total international trade (UNCTAD, 2004a). The impact of ICTs on the performance and competitiveness is achieved through increased information flows, which result in knowledge transfer as well as improved organization. The application of ICTs in business ranges from Computer-Aided Design (CAD), Computer-Aided Engineering (CAE) and Computer-Aided Manufacturing (CAM) to Intranet, Internet and Electronic Commerce (E-Commerce) among others. 'E-Commerce is one of the most visible examples of the way in which ICTs can contribute to economic growth. It helps to integrate economies with the global economy. It allows business and entrepreneurs to become more competitive. And it provided jobs, thereby creating wealth' (UNCTAD, 2004a).

International competitiveness in trade, as the chapter on Nigeria points out, continues unabated in spite of the growing number of supra-national organizations such as the European Union (EU), North American Free Trade Zone, Southern Common Market *or Mercado Común del Sur* in Spanish (MERCOSUR), Economic Community of West African States (ECOWAS), and the Southern African Development Community (SADC). While African countries are widely regarded as being non-competitive, there are a few success stories. Some garment-producing firms in Mauritius have gained prominence as exporters even to the EU and United States markets since the early 1980s. UNCTAD (2004b) notes that since the USA's Africa Growth and Opportunity Act (AGOA) came into effect in 2000, many export-oriented garment factories have been set up in Namibia, Mozambique, Swaziland, Lesotho as well as Mauritius. The report went further to say that Lesotho became the largest African apparel exporter to the USA in 2003, with about 10,000 newly created jobs. Also, Nnewi industrial cluster in Nigeria is another success story, even though the manufactures especially automotive spare parts are exported to other West African countries.

Improving business competitiveness requires a variety of systemic transformation, including the availability of efficient infrastructure and services, technological changes, new organizational processes, adoption of good practices, training, mobilization of underutilized resources, creation and segmentation of markets, product and process certification and efficiency in inputs utilization. A study of small and medium enterprises in OECD (1997) on the challenges of globalization reveals that, internal dynamics of firms and environmental support are key factors which could either facilitate or inhibit export. SMEs that are active players in the global economy make special efforts to search for diversified growth by pursuing innovation-based production, and open-minded management capable of engaging the appropriate specialized resources e.g. ICTs.

The enabling environment includes effective consulting, funding and logistical resources to support exports. Some of the key internal factors inhibiting the globalization of SMEs are lack of experience on the part of the management, inadequate resources and an excessive risk perception. Poor national information networks with weak or inadequate international connection, poor regional resources and support programmes are also responsible for this (Julien *et al.*, 1993). Globalization and competitiveness of SMEs and Large-Scale Enterprises (LSEs) are driven by similar forces; the pre-eminence of ICTs in this regard is unassailable. A critical question for developing countries is whether the adoption of ICTs can be an authentic vehicle for solving the problems of economic development and low level participation in international trade.

6   The role of ICTs

Realizing the potential of ICTs in electronic business (e-business), as the chapter on India points out lucidly, the United Nations Commission on

International Trade Law (UNCITL) adopted a Model Law in 1996. The UN General Assembly recommended to its members in January 1997 that they give due consideration to this Model Law, when they enact their laws related to e-business. Despite several advantages, the growth of e-business has been dismal particularly in developing countries. The major impediments in the adoption of such technologies are the validity and authenticity of information. Lack of proper cyber laws could also be held responsible for the slow changeover to e-business.

There are many descriptions of e-business. In essence, e-business is about business innovation, about serving new and changing markets. E-business is meant to reshape the way companies go to markets, the way customers buy products and services. It can also be defined as a tool that forward-looking enterprises are racing to adopt. E-business technologies are meant to help adopters to reach new customers more efficiently and effectively. E-business transforms the exchange of goods, services, information, and knowledge through the use of ICTs. There are several models of e-business, namely, (i) business to business (B2B), (ii) business to consumer (B2C), (iii) consumer to consumer (C2C), (iv) business to government (B2G), and government to business (G2B).

Broadly speaking, there are three modes of e-business transactions. These are offline, online, and e-business using shared or individual portals. First and comparatively less effective than other forms of e-business tools, that is, offline e-business is enabled by electronic messaging systems. Offline e-business is normally done through email systems while online e-business transactions take place with Uniform Resource Locators (URLs) of companies. Having a URL does not necessarily mean that an enterprise is able to process online e-business transactions. URL must be dynamic and should have online transaction facilities such as Active Server Pages (ASPs) that allow online transactions. The third and most effective way of doing e-business is through portals. Portals are the essential additions in network technologies. They fulfil an important role of aggregating contents, services, and information on the net. Broadly speaking, their position on the net is between users (buyers) and web contents. This unique position enables portals to leverage marketing and referrals as they are intermediaries between web users and companies.

According to the information supplied by the International Data Corporation (IDC), the Compound Annual Growth Rate (CAGR) of e-business in the global market has been 105 per cent during 1995 and 2000. E-business has emerged as the fastest growing technology in recent times. Although there are several models of e-business, only two models, namely, B2B and B2C have experienced the highest growth. Within these two, B2B has grown from 3.5 billion US$ in 1995 to 34.0 billion US$ in 1998 while the growth of B2C has been 1.0 billion US$ in 1995 to 4.0 billion US$ in 1998. The share of B2B has increased from 77.78 per cent in 1995 to 89.47 per cent in 1998. These data suggest that B2B has grown faster than B2C worldwide.

7 Finding from country studies

The study on India aims at identifying the factors that have significant impact on the diffusion of e-business, assessing the contributions of existing ICT industry in changeover to e-business, the role of institutional environment in growth of e-business and the impact of e-business on transaction and co-ordination costs.

The chapter analyses the determinants of the adoption of e-business technologies in the manufacturing sector in India. Based on the survey conducted during December 2000 and February 2001 the analysis refers to 51 firms located in New Okhla Industrial Development Area (NOIDA). Firms were revisited during January to March 2005 and found that technological profile of firms did not change much. Sample firms are dominated by SMEs as 73 per cent of these firms employ less than 150 persons. Entrepreneurial characteristics such as managing director's education and age, historical data of firms, and other firm-specific factors such as size of operation, export intensity, international orientation, wage rates, profit margins were included in the analysis. It also included a variable, i.e. bandwidth that represents the institutional environment created by central and local governments. The opinions of managing director (MDs) on potential benefits of the adoption of e-business technologies were also considered.

The analysis has been carried out by grouping the sample firms into three categories, namely, offline, online, and portal-using firms. Firms doing e-business using offline technologies such as messaging systems were grouped in the first category, whereas firms that were using messaging systems as well as online e-business tools such as ASPs in their URL were classified as online e-business doing firms. Portal and ERP using firms were considered as advanced users of e-business technologies and were treated in the third category. Firms were assigned ranks depending on the type of e-business technology used by them. As far as the model of e-business is concerned, 92 per cent of the sample firms adopted B2B e-business model.

The multivariate ordered PROBIT technique used to identify the determinants of the degree of the adoption of e-business tools reveals that the firms with more international orientation adopted more advanced e-business tools. Wage rates and scale of operations also emerged as significant determinants of the adoption of e-business technologies. The study captures the role of bandwidth in diffusion of e-business technologies. The findings tend to support several other studies pursued in the past (Doms *et al.*, 1997; Bedi, 1999; Siddharthan, 1992; Cohen, 1995; Pavitt *et al.*, 1987; Soete, 1997).

A positive association between type of e-business technology used by firms and the bandwidth is very much evident. Earlier a study by

NASSCOM (2000) suggested that availability of higher bandwidth is a prerequisite for the penetration of Internet and web-enabled services in India. The present study concludes that a very reliable and affordable telecommunication network has to be in place to harness the potentials of ICTs. The IT Law 2000 is a necessary but not sufficient condition for the success of diffusion of ICTs. The initiatives towards communication technology convergence are expected to result in faster growth of e-business. There is indeed a need to create proper local, national, and global information infrastructure to derive the maximum benefit from the ICT revolution.

The chapter on Malaysia showed that the majority of the respondents had advanced ICT adoption rate though generalization may not be justified because of the fact that majority of the sample were drawn from more advanced or technology driven industries such as hardware and machinery, chemical and pharmaceutical and electronics. One important finding is that educational attainment of companies' directors had a bearing on ICT adoption. There exists a strong link between technology driven firms and the qualification of their leaders. Collaboration with foreign partners was found to be a usual practice among SMEs in Malaysia. Language barrier was seen as the main problem, hindering training or collaborative efforts with foreign partners.

The findings suggest that other than providing financial support to SMEs, the Malaysian government need to focus on human resource development policies. Technological institutions that can provide job-oriented formal training and short-time training opportunities near the workplaces may be helpful for skill up-gradation of workers. A provision of such opportunities is expected to contribute to efficiency and productivity of workers. However, the study could not evaluate the impact of ICT on economic performance of sample SMEs due to the lack of data on performance indicators.

The study on Costa Rica can be summarized in three parts. First it evaluated economic performance of Costa Rican SMEs and their contributions to the national economy. Despite their substantial contributions, they were not given due importance until recently. The second part of the study delineates the policy initiatives that were taken since mid-1980s and were exclusively meant for SMEs. The third part of the study focuses on the innovativeness of SMEs in general and identification of factors that resulted in varying degree of ICT adoption based on a survey of 68 firms conducted during July 2004 and February 2005.

Although SMEs contribute significantly to the national economy in terms of jobs and exports, public efforts to improve their competitiveness have been erratic or weak. Hence, the progress towards that objective seems to be nominal. In terms of industrial policy, the experience of the last 20 years suggests that an overall industrial strategy should be developed with a long run perspective. SMEs are relatively young companies managed by middle aged, highly educated managers. Common strategies to compete are quality and product differentiation in specific niches. Training is important but R&D has very few rooms in the SMEs agenda. ICTs play a highly significant role in the performance of SMEs. Although the country has been able to provide extensive coverage of basic technologies in the last 40 years (fix lines, for instance), it also lags behind in terms of modern technologies such as Internet coverage and speed.

Although the package of Internet services provided by National Institute of Electricity allows firms to use basic functions, lack of an adequate infrastructure is impeding SMEs in using the Internet for productive, marketing and sales purposes. In general, improved performance in productivity, time advantages, design flexibility, reorganization and management are associated with an extensive use of advanced ICTs, supporting the view that ICTs are essential for enhancing the competitive profile of the firm.

While analysing the data in multivariate analysis framework firms were categorized into three groups, namely: low level of ICT using firms, moderate users of ICT, and advanced ICT using firms. The ICTs[1] that were included in the analysis are email, Internet, portal, web enabled technologies, MIS, CAD/CAM, CAE, FMS, and CNC. Clustering of firms based on their intensity of ICT use was imperative as the sample had to be divided into reasonable number of distinctive groups. Subsequently, discriminant analysis was carried out to identify factors that discriminated different levels of ICT using firms.

Factors representing, entrepreneurship, international orientation, causes and consequences of ICT use, sources of competitiveness, and knowledge acquisition opportunities were used in the analysis. The results suggest that general managers (GMs) knowledge base and academic background emerged as an important discriminant. The study finds that skill intensity of advanced ICT using firm was higher than the rest. International orientation of firms also discriminated three groups of firms. GMs that attributed higher importance to ability of ICTs in providing better management control and useful market information adopted more advanced ICTs. On the whole, the adoption of ICTs results in productivity gains and efficiency in production processes. The government needs to encourage and provide institutional support for SMEs to participate in international markets so that they remain globally competitive. Lack of this could be attributed to the present state of low level of globalization of Costa Rican SMEs. Greater participation in global market might enable firms to increase their contribution to the national economy.

---

1 Management Information System (MIS), Computer-Aided-Design/Computer-Aided-Manufacturing (CAD/CAM), Computer-Aided-Engineering (CAE), Flexible Manufacturing System (FMS), and Computerised Numerically Controlled (CNC).

The study on Jamaica based on the primary data collected from 60 SMEs located in the Greater Kingston Metropolitan area focused on technology profile and international orientation of the units. ICT tools covered in the study include Internet, email, web enabled technologies, MIS, CAD/CAM, CAE, FMS, and CNC. Findings of the study suggest that most of the firms were using one or the other type of ICT. While Internet was preferred technology in the non-production processes CAD/CAM was popular in production processes.

Based on the factor analysis a composite score of ICT intensity used by sample firms was estimated. The production function approach was used to identify the role of ICTs in augmenting performance of firms. The findings suggest that the performance, represented by sales turnover was significantly influenced by intensity of ICTs used, skill intensity of workforce, foreign collaboration, age of owners/managing directors, size of firm, formal training as a mode of knowledge acquisition, and communication technology infrastructure represented by the speed of communication.

The study finds evidence in favour of the critical role played by the age of managing directors. Although, age and academic qualification have positive association, the age has been able to capture the type of education provided in recent times. For instance the type of training provided to an engineering graduate 20 years ago is very different from today. In recent times due emphasis has been given to the application and potential benefits of ICTs in all training and educational programmes. Consequently younger managing directors with the same qualifications as older ones tend to adopt more advanced ICTs. The study also finds evidence to support the argument that formal training of workers is preferred to learning by doing mode of knowledge acquisition. Managing directors' preference for formally trained workers in influencing size of operation is a case in point. The other findings lend support to several earlier studies (Lal, 2004; Earl, 1989; Doms *et al.*, 1997).

Policy implications of the study are twofold: one, Jamaican government needs to improve accessibility of telecommunication network at a globally competitive rate. That in turn is expected in higher adoption of ICTs, which might influence performance of firms in the domestic market. Government also needs to provide marketing support to enable SMEs to venture into international markets. Greater integration of SMEs with the foreign market might contribute to export growth. Another important implication of the study is related to the mode of knowledge acquisition. Government needs to strengthen the existing technical training institutions so that they can produce persons with appropriate skills needed by SMEs.

The study on Nigeria examines the following two questions: (a) what strategies have proved useful in terms of enabling Nigerian SMEs to become more competitive through the use of ICTs? And (b) how can ICTs be used to facilitate the participation of Nigerian SMEs in national and international supply chains? The survey questionnaires retrieved from 67 SMEs in Aba, Ibadan, Lagos and Nnewi were analysed. Only 12 of the sample firms did not adopt any of the ICTs while 30 and 25 were classified as moderate and advanced adopters respectively based on the number of ICTs used. Overall 3,195 persons were employed by the 67 sample firms with the following classification: non-adopters 647 (20.25 per cent), moderate adopters 804 (28.26 per cent) and advanced adopters 1,645 (51.49 per cent). About 9 per cent of the entire work force had received one form of ICT training or the other, and the number of such trained workers increased with the level of adoption.

About half of the 67 sample firms had one form of collaboration or the other while roughly 12 per cent had to pursue their business without any foreign partnership. Among the non-adopters one firm was engaged in collaboration with foreign partners, whereas 68 per cent and 50 per cent of the advanced and moderate ICT adopters respectively had foreign partners. The two main preferred modes of technology transfer were importation of machinery and importation of technical design and drawings. Payments for the technologies acquired were usually in the form of royalties followed by technical fees and lump sum payment. Some of the major benefits emanating from foreign partnerships include access to foreign technology, goodwill, marketing, brand name and international linkage. Some of the SMEs depended on certain raw materials, machines, equipment and technical support from foreign countries especially the EU, and USA. On the other hand, some local firms exported raw materials, such as cocoa beans and timber, intermediate products (e.g. cocoa cake) and finished products like apparels, plastic products, biscuits, computer parts and software. This finding lends credence to the fact that successful firms under globalization need to develop and sustain domestic and international networks and partnerships.

In general, firms (44.77 per cent) rated the role of ICTs in foreign collaboration very high. Only 13.43 per cent of all the firms either rated ICT as being not important at all or unimportant while 41.79 per cent were indecisive. Changes in product mix through diversification and or acquisition of firm(s) producing different product lines represent progress and advances in technological capability and response to competition. Four out of every ten responding firms claimed to have made average changes in their product mix within the preceding 10 years. Nonetheless, the positive influence of ICT adoption on product mix changes is confirmed by the higher proportion of advanced adopters, which claimed to have made many changes in their product mix compared to moderate adopters and non-adopters. With respect to competitiveness, the firms intend to meet this challenge by improving the quality of their products, flexibility in design, protection of market share, and goodwill/brand name among others.

SMEs development, growth and competitiveness under rapid globalization are contingent upon unhindered access to information about the local and international operating environments and challenges. However, the absence of business development services has made it impossible for gathering and disseminating such information. There is, therefore, an urgent need for the establishment of business development service centres in different parts of the country within the framework of Small and Medium Enterprises Development Agency of Nigeria (SMEDAN), and relevant private sector stakeholders. The centres should serve to educate SMEs on benefits of appropriate technologies and technical partners and facilitate access to them. Information about market demand and supply by sector and geographical location is also very critical to the competitiveness of firms. Availability of these types of information would most likely prompt them to look beyond central Africa and ECOWAS as potential export markets for their goods and services. In this regard, SMEDAN and the business development service centres, organized private sector associations and trade sections of Nigerian embassies have appropriate roles to play.

One of the major factors inhibiting ICT diffusion and intensive utilization is poor physical infrastructure such as adequate and uninterrupted electricity supply, and communication connectivity infrastructure. The resultant effect is high cost of ICT acquisition and maintenance. The government should urgently spur existing public-private partnerships (the private telephone operators) with a view to improving the state of these types of infrastructure. The Central Bank of Nigeria should in partnership with SMEDAN and relevant private sector stakeholders, organize seminars with a view to sensitizing SMEs on the benefits and modalities of accessing Small and Medium Enterprises Equity Investment Scheme (SMEEIS) fund for sustainable growth and ICTs acquisition and utilization. Since ICT is a major driver of the globalization process, for Nigerian SMEs to be competitive they must have reasonable level of competence in ICT applications. Consequently, trade and business associations with the active support of relevant government agencies should organize regular training workshops on the various facets of e-business.

Furthermore, the efficiency of business decisions is known to be contingent on the ability of managers to assess potential opportunities and risks, and to identify and select most appropriate actions to pursue business objectives subject to external and internal constraints. There should therefore be deliberate entrepreneur skills improvement programmes for SMEs. Under the current globalization, it is no longer only individual firms that are engaged in competition rather groups of firms organized in networks, whose dynamic development and competitiveness depend on the spatial location and proximity (real and virtual) with R&D facilities, technology formation and dissemination institutions, knowledge centres, financial and export information institutions. Consequently, the government should provide an enabling environment for SMEs to operate in efficient ways and access to relevant research and development results and market information so as to make them competitive.

The global ICT revolution offers Nigeria the unique opportunity of actively participating in a globalized economy. The biggest beneficiaries are countries that are proactive and have identified the strategic relevance of ICTs in the rapid transformation of national economic development. A major policy challenge therefore is how the country could re-engineer and re-position its formal and informal educational and training systems in order to rapidly raise ICT literacy level. In specific reference to SMEs, ICT training sessions may be organized in or around the clusters they are located in.

2

E-business and Manufacturing Sector: A Study of Small and Medium-sized Enterprises in India[2]

1 Introduction

It has been argued in the literature that the adoption of Information and Communication Technologies (ICTs) allows a reduction in co-ordination costs and leads to efficient electronic markets (Malone *et al.*, 1987; Lee and Clark, 1997). Several studies in Pohjola (2001) have found significant returns on ICT investments in developing as well as developed countries. A partial survey of literature by Bedi (1999) on role of ICTs in economic development suggests numerous benefits of its adoption. These benefits range from employment, to productivity gains, consumer surplus, and improvement in product quality, etc. In fact several nations started searching for alternatives to paper-based methods of communication and storage of information in the last quarter of the twentieth century. Realizing the potential of ICTs in electronic business (e-business), the United Nations Commission on International Trade Law (UNCITL) adopted a Model Law in 1996. The UN General Assembly recommended to its members in January 1997 that they give due consideration to this Model Law, when they enact their laws related to e-business. Despite several advantages, the growth of e-business has been dismal particularly in developing countries. The major impediments in the adoption of such technologies have been the validity and authenticity of information. Lack of proper cyber laws could also be held responsible for the slow changeover to e-business.

There are many descriptions of e-business. In essence, e-business is about business innovation, about serving new and changing markets. E-business is meant to reshape the way companies go to markets, the way customers buy products and services. It can also be defined as a tool that forward-looking enterprises are racing to adopt. E-business technologies are meant

---

2 This chapter has already been published in *Research Policy*, 2002.

to help adopters to reach new customers more efficiently and effectively. E-business transforms the exchange of goods, services, information, and knowledge through the use of ICTs. There are several models of e-business, namely, (i) business to business (B2B), (ii) business to consumer (B2C), (iii) consumer to consumer (C2C), (iv) business to government (B2G), and government to business (G2B).

Broadly defined, there are three modes of e-business transactions. These are offline, online, and e-business using shared or individual portals. First and comparatively less effective than other forms of e-business tools, that is, offline e-business is enabled by electronic messaging systems. Offline e-business is normally done through email systems while online e-business transactions take place with Uniform Resource Locators (URLs) of companies. Having a URL does not necessarily mean that an enterprise is able to process online e-business transactions. URL must be dynamic and should have online transaction facilities such as Active Server Pages (ASPs) that allow online transactions. The third and most effective way of doing e-business is through portals. Portals are the essential additions in network technologies. They fulfil an important role of aggregating contents, services, and information on the net. Broadly speaking, their position on the net is between users (buyers) and web contents. This unique position enables portals to leverage marketing and referrals as they are intermediaries between web users and companies.

According to International Data Corporation (IDC) data sources, the Compound Annual Growth Rate (CAGR) of e-business in the global market has been 105 per cent since 1995. E-business has emerged as the fastest growing technology in recent times. Although there are several models of e-business, only two models, namely, B2B and B2C have experienced the highest growth. Within these two, B2B has grown from 3.5 billion US$ in 1995 to 34.0 billion US$ in 1998 while the growth of B2C has been 1.0 billion US$ in 1995 to 4.0 billion US$ in 1998. The share of B2B has increased from 77.78 per cent in 1995 to 89.47 per cent in 1998. These data suggest that B2B has grown faster than B2C worldwide.

In Indian context, the government had taken initiative since late 1980s in the adoption of e-business by allowing companies to file their annual returns through floppies. However, the diffusion of e-business could not gain momentum due to the reasons mentioned earlier. Having taken cognisance of the UN General Assembly resolution and the recommendations of Indian industry associations, the Government of India drafted an IT bill to boost e-business in the country in 1999. Passed by the Parliament on May 16, 2000, the IT Bill 99 deals with the matters such as digital signature, electronic governance, electronic acknowledgement, acquisition and despatch of electronic records, regulation of certifying authorities, penalties and adjudication, the cyber regulations appellate tribunal, offences, and laws related to network service providers.

With the passage of the IT Bill 99 by the Parliament, e-business and e-governance are likely to grow as fast, if not faster, than the ICT sector itself. However, there are reasons to be skeptical about the diffusion of e-business. The e-commerce scenario in India is more or less similar to the global picture. As far as the share of B2B to the total e-business is concerned, it is static having maintained a constant share (90 per cent) since 1998. According to an estimate by NASSCOM (2000), the growth of e-business is expected to be the same, if not more, than in the global market.

Several aspects are associated with the adoption of e-business. First, the identification of potential areas in which e-business technologies can be adopted. Second, the availability of telecommunication networks. Third, the potential gains expected from e-business. And fourth, the impediments associated with e-business in a developing context. It may not be possible to cover all these aspects in one study.

The main objectives of this study however are:

- The identification and analysis of factors that have significant impact on the diffusion of e-business.
- An examination of the contributions of existing ICT industry in change-over to e-business.
- Role of institutional environment in growth of e-business.
- An assessment of the impact of e-business on transaction and coordination costs.
- An evaluation of the prospects of e-business in India.

The chapter is organized as follows: The analytical framework is discussed in Section 2. The data and methodology used for identifying the factors that were associated with the adoption of e-business are discussed in Section 3 whereas the hypotheses are formulated in Section 4. Section 5 presents the statistical results while the findings of the study are summarized in Section 6.

2   Analytical framework

The diffusion of any new technology depends on several factors such as the potential benefits of technology, absorptive capacity of firms, and the institutional environment prevalent in the country. Potential benefits need not necessarily be related only to manufacturers of products and services that use new technologies but may affect consumers of products and services as well. E-business technologies, the latest development in ICTs have potentials to benefit not only producers but also users of services and products. The explosive growth of Internet is a case in point. The worldwide number of Internet users have increased from 16 million in 1995 to 250 million in 2000 (NASSCOM, 2000).

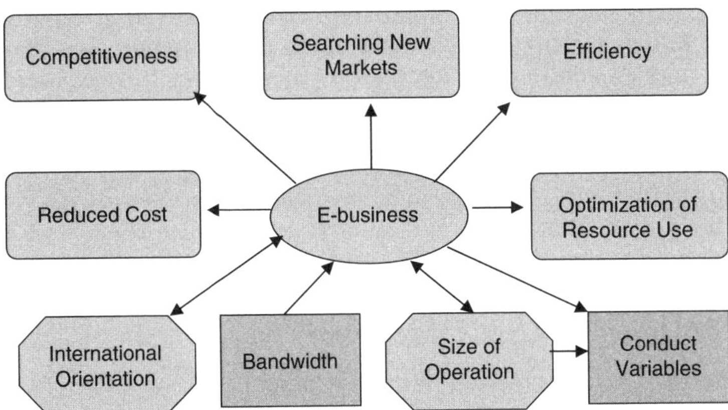

Figure 2.1   Firms' characteristics and e-business linkages

Adoption of e-business technologies is a function of several factors. Many times these factors mutually reinforce each other. Interaction between the causes and consequences of adoption of e-business technologies is depicted in Figure 2.1.

It can be seen from Figure 2.1 that international orientation, which is represented by imports, exports, technological and financial collaboration and the adoption of e-business mutually reinforce each other. This is because the use of ICTs in general results in the death of distance (Soete, 1997). Moreover, e-business is expected to result in fundamental changes in the market landscape. Similarly, sales turnover and the adoption of e-business are expected to influence each other. This holds true not only for adoption of e-business but for any innovative activity carried out on the shop floor.

One of the major pre-requisites for the success of e-business is the existence of a very strong and reliable communication network. Access to a higher bandwidth is not within the control of individual firms. It forms part of the institutional infrastructure provided by governments. Realizing the important role that availability of higher bandwidth plays in the success of web-enabled services, greater emphasis has been laid on the privatizing and de-regulating of telecommunication services in developing and developed countries. Conduct variables such as skill intensity of firms, investment on R&D, wage rates are important factors that are expected to have bearings on the adoption of e-business. Higher remuneration is always a major incentive for the workforce to create and adopt innovations effectively and efficiently.

Numerous benefits that flow from the adoption of e-business have been cited in the literature (Gray and Prahalad, 1994). Many firms adopt e-business because of low coordination costs. However, as depicted in Figure 2.1, many more benefits exist. Due to the inexpensive access to global markets and information that Internet enables, it is fast becoming the world's largest and most versatile marketplace for services, products, and information. E-business has the potential to redefine the existing business infrastructure organizations and to re-evaluate the way in which they do business. It has capabilities in re-engineering business processes across the boundaries that have traditionally separated suppliers from their customers. Previously separated activities such as order processing, payments, and after sales services may be merged into a single process. This results in reducing the costs of creating, moving, processing, and managing documents.

E-business is also expected to reduce operational costs since electronic information tends to be more accurate, and timelier and easily available. Another benefit of e-business could be the higher efficiency obtained in business transactions due to a fast and accurate processing of information. Web-enabled services are likely to strengthen the competitiveness of firms as these technologies may change the relationship with customers by creating a stronger link between firms and its clients.

3 Data and methodology

The study is based on primary data collected from firms located in the New Okhla Industrial Development Area (NOIDA). This is a newly developed industrial town near the national capital, New Delhi. There are four main ISPs (Internet Service Providers), namely, Videsh Sanchar Nigam Limited (VSNL), Software Technology Park of India (STPI), Satyam, and Mantraonline operating in NOIDA. Mantraonline and Satyam are private sector ISPs whereas VSNL and STPI are state-owned public sector companies that are involved in providing Internet services. In addition, M/s India Markets is another private sector e-business solution providing company located in NOIDA. We approached all ISPs and India Markets to get information about the firms that have adopted e-business technologies in NOIDA. We could obtain a list of 68 firms, which included the name of managing director (MD) and address of firm. Product profile of sample firms is presented in Appendix Table 2.1.

It may be safely assumed that all the firms that are using some form of e-business model and are located in NOIDA have been covered in this study. We approached all 68 firms during the survey. However, we could get data only from 51 firms, a response rate of 75 per cent. The survey was conducted during December 2000 and February 2001. Historical, financial, technological, and data on international orientation of firms were collected through a semi-structured questionnaire. Historical data include the background of MD and age of firm whereas financial data (1999–2000) consist of sales turnover, investment on ICTs, wage bill, exports, imports, profit after tax, and value added, etc. Technological data include the type of e-business tool adopted and bandwidth being used by firms. Data on technological and financial collaborations with multinational companies, and opinion of MDs on potential benefits of e-commerce were also collected. All the opinion variables were measured on a five-point scale. Firms were revisited during January to March 2005 and found that technological profile of sample firms did not change much.

The relationship between the degree of e-business adoption and the other variables is likely to be:

EB_TYPE = {Entrepreneurial characteristics, International orientations, Size of operation, Wage rate, Bandwidth}

3.1   Independent variable

Type of E-business Technology Adopted by Firms (EB_TYPE):

Firms using offline e-business technologies were assigned rank 1 whereas the online technology using firms have been assigned a higher rank that is 2, because all the second category of firms possessed an email facility also. The third category of firms were assigned the highest rank, that is, 3 as the portals allow offline and online e-business also. Hence the independent variable, that is, the type of e-business technology adopted is not only discrete but is an ordered variable.

As the independent variable, that is, type of e-business tool adopted by the firms is a discrete and ranked variable, it was considered appropriate to apply the discrete dependent regression model to identify the determinants of the adoption of intensity of e-business tools. We found that the multinomial ordered PROBIT model could be the most appropriate technique that can be used to analyse this kind of data. Four equations using the PROBIT model were estimated. We also analysed the data in a univariate framework. The univariate and multivariate statistical results are presented and discussed in Section 5.

4   Hypotheses

Drawing upon the theoretical and empirical evidence on the adoption of ICTs and the consequences thereof, we now proceed to formulate the hypotheses concerning the nature of firms that have adopted e-business technologies in different degrees. All the variables that have been included in this study can be grouped in two categories: First set of variables is represented here by MD's education, wage rate, sales turnover, exports, bandwidth, collaboration, and profit margins while the second set of variables consist of opinion variables.

All the sample firms have been grouped into three categories depending on the type of e-business technologies adopted by them. The first category of firms were doing e-business through offline technologies. Firms that were using electronic mail systems for business activities have been categorized as firms using offline e-business technology. During the survey, it was found that many firms have their presence on the net. They have their URLs with some degree of online e-business facilities such as ASPs. All those firms that have dynamic URLs have been categorized as online e-business doing firms. The third category of firms were those that were doing e-business using the latest technology, that is, portal-based e-business.

The hypotheses concerning all the variables included in the analysis are discussed later in detail.

4.1 Managing directors' education (MDEDU)

The actual data on entrepreneurs' qualification were collected and this education was converted on a quantitative scale. Graduate entrepreneurs are thus assigned the lowest rank, that is, 1 whereas the postgraduate (PG) MDs were given a rank higher than the graduates, that is, 2. It was found during the survey that a large number of entrepreneurs were engineering graduates. These were separated from the ordinary graduates and treated as more qualified than ordinary postgraduates. Entrepreneurs with bachelor of engineering (BE) degree have thus been given rank 3. There are many entrepreneurs who have either a master's degree in engineering or a professional degree such as master of business administration (MBA). All such entrepreneurs have been assigned highest rank, that is, 4. The distribution of firms according to entrepreneurs' qualification and type of e-business technology adopted by firms is presented in Table 2.1.

It can be seen from Table 2.1 that the percentage of graduate entrepreneurs decrease as the intensity of e-business adoption increases from offline to portal. Whereas the percentage of MDs, who have highest technical qualifications, increases from 12 per cent in offline e-business using firms to 63.64 per cent in firms that were doing business through their own or shared portals. This indicates that the firms that are managed by technically qualified persons have adopted a higher degree of e-business technologies.

The importance of entrepreneurship has been given a pivotal place in Schumpeterian and neo-Schumpeterian literature of technological change. Empirically, the role of entrepreneurs' qualification has been analysed by many scholars. The findings of several studies (Earl, 1989; Cohen, 1995; Lal, 1996) suggest that entrepreneurs' knowledge and qualification play a very important role in influencing the degree of adoption of ICTs. Drawing upon the theoretical literature and findings of the several empirical studies, we hypothesize that the entrepreneur's qualification is likely to emerge as a

Table 2.1 Entrepreneur's qualification and adoption of e-business

| | Qualification of entrepreneurs | | | | | | | | |
EB-type	Graduate	%	PG	%	BE	%	BE and MBA	%	Total
Offline	4	16.00	9	36.00	9	36.00	3	12.00	25
Online	1	6.67	3	20.00	7	46.67	4	26.67	15
Portal	1	9.09			3	27.27	7	63.64	11
Total	6	11.76	12	23.53	19	37.25	14	27.45	51

significant factor in influencing the type of e-business technology adopted by firms.

4.2 Technological collaboration (COL_TECH)

The distribution of firms according to technological collaboration with foreign firms and the type of e-business technology adopted by firms is presented in Table 2.2.

Table 2.2 shows that the majority of firms (72.73 per cent) that have technological collaboration with foreign firms are doing e-business through portals while 68 per cent of firms that are using offline techniques of e-business do not have foreign collaboration. This indicates that the type of e-business tool used by the firms and the technological collaboration of foreign firms are related to each other.

Several scholars (Stiglitz, 1989; Evans and Wurster, 1997) have emphasized that ICTs play an important role in exchanging information, knowledge, and product designs between manufacturers and suppliers of technology. One of the major contributions of ICTs in the business environment is the better co-ordination of manufacturing activities. Online e-business tools may be the best-suited technology to co-ordinate with foreign companies particularly. Hence a positive relationship between type

Table 2.2 Technological collaboration and adoption of e-business

| | Technological collaboration | | | | |
EB-type	Yes	%	No	%	Total
Offline	8	32.00	17	68.00	25
Online	10	66.67	5	33.33	15
Portal	8	72.73	3	27.27	11
Total	26	50.98	25	49.02	51

Table 2.3   Bandwidth and adoption of e-business

			Bandwidth				
EB-type	**Telephone**	**%**	**64 kbps**	**%**	**128 kbps**	**%**	**Total**
Offline	23	92.00	2	8.00			25
Online	11	73.33	4	26.67			15
Portal	1	9.09	6	54.55	4	36.36	11
Total	35	68.63	12	23.53	4	7.84	51

of e-business tools adopted by firms and technological collaboration is hypothesized.

4.3   Bandwidth (BANDW)

Data on actual bandwidth that was being used by firms were collected. However, the values of the variable have been converted on a new scale for the purpose of analysis. The telephone-using firms are given value 1 while 64 kbps (kilo bits per second) ISDN (Integrated Service Digital Network) line-using firms are assigned value 2. The firms that have access to 128 kbps bandwidth are assigned value 3. Various type of communication modes such as, telephone line, ISDN line, Radio link, and VSAT (Very Small Aperture Terminal) are available in India. However, during the survey it was found that firms were either using usual telephone lines or ISDN lines. Within the digital category, firms were using dialup ISDN line of 64 kpbs or 128 kbps. There was only one firm that was connected with its foreign partner in Japan with an ISDN leased line. Table 2.3 shows the distribution of firms according to use of bandwidth.

It can be seen from Table 2.3 that a large number of firms (92 per cent) that have adopted offline technologies are using a simple telephone line for Internet connectivity while all firms using 128 kbps lines are those that are doing e-business through portals. Although a few firms engaged in either offline or online e-business have also preferred 64 kbps digital line, a majority of firms (90.91 per cent) that are portal-using firms have adopted the digital mode of communication.

Empirically, several studies have found that communication networks are pre-requisites for the success of the ICT revolution. The findings of the World Bank Report (1998) and Bayes and Braun (1999) are related to rural communication in Costa Rica and Bangladesh, respectively whereas NASSCOM's study (2000) covers the various industrial sectors in India. NASSCOM (2000) suggests that not only reliable and high speed communication systems but also the existence of a cheap communication network is essential for the success of e-business. In view of these empirical evidences, it is hypothesized that there exists a positive relationship between bandwidth and the type of e-business tool adopted by sample firms.

4.4   Ratio of total expenditure on salary to total employment (WRATE)

The wage rate has been computed as the ratio of total wages paid per month to the total employees of a firm. Although firms do not require very highly paid workers for use of e-business tools, the average wage paid to an employee by portal-using firms is expected to be high. This is because the MDs of portal-using firms are technology savvy and they are aware of the fact that e-business is a tool for business innovation. E-business alone cannot improve the performance of firms. E-business can play the role of a catalyst in achieving better performance along with other factors such as quality of products, after sales support, and innovative capabilities of firms. In order to manufacture products of flexible design and an improved quality of products, the firm need to employ skilled and experienced workers. The findings of a study by Lal (1996) suggest that the average wage paid to an employee by advanced users of IT tools were higher than firms that were non_IT using firms. We also expect that there exist a similar relationship between wage rates and the type of e-business tool used by firms.

4.5   Firm size (STO)

Sales turnover in Rs. million has been used as a proxy of size of firms. A positive relationship is generally postulated between size and technological progress of firms in the Schumpeterian and neo-Schumpeterian literature. It is the size of a firm that provides the financial resources to acquire new technology and enables spending on innovative activities. This is particularly true in the case of SMEs where firms do not have an easy access to financial institutions. Moreover, SMEs have a tendency to spend an affordable amount from their own resources on innovative activities. Several studies (Siddharthan, 1992; Lall, 1983) found a positive relationship between innovative activities and size of operation of firms. We also postulate a similar relationship between e-business technologies adopted by firms and size of operation.

4.6   Exports intensity (EXPOINT)

Exports intensity has been computed as the ratio of total exports to sales turnover. Conflicting findings with respect to exports and adoption on advanced technologies have been reported in the literature. Several reasons have been cited for conflicting results. It is argued in the literature that firms do not always adopt advanced technologies to meet challenges in the international markets. If the domestic market is not protected, it becomes imperative for firms to upgrade technology for their survival in the domestic market. In that case technological advancement may not be related to the export intensity of firms. Although the Indian market is no more a protected market, the entry of foreign firms in the SMEs sector is not very high. Firms that are operating in the domestic market may not be facing quality competition as severe as firms that are operating in international markets. Moreover, online and portal-based e-business technology may be more beneficial to export-oriented firms. Hence, we hypothesize that e-business technologies are positively related to the export performance of firms.

4.7   Profit margins (PRM)

The variable has been computed as the ratio of profit-after-tax to sales turnover. It is expected that the use of e-business technologies are likely to reduce costs in activities other than manufacturing. This in turn should be reflected in higher profit margins. It is more likely to be true for SMEs where firms do not invest in any new technology unless they anticipate immediate returns on investments. Unlike large companies, SMEs prefer to adopt technologies that have a lesser gestation period. In view of the fact that the sample firms in this study are SMEs, we expect that higher profit margin is driving force behind the adoption of e-business technologies.

4.8   Opinion variables

E-business technologies promise numerous benefits such as better customer relation management (CRM), improved supply chain management (SCM), reduction is errors and costs, optimization of resource use, searching new markets, efficiency in business transaction, and augmentation in the competitiveness. We have tried to capture the role of perception of MDs about e-business technologies through their opinion on the following two variables that are considered to be more relevant for sample firms. These variables have been measured on a five-point scale. MDs responses are discussed in the following sub-section.

4.8.1   Efficiency in business transactions (EFFITRANS)

The first opinion variable is the 'efficiency in business transaction'. The perception of MDs on this variable is presented in Table 2.4.

It can be seen from Table 2.4 that 25.5 per cent of sample firms did not consider any significant impact of e-business tools on the efficiency of

Table 2.4   Efficiency in business transactions

EB-type	Not significant	Significant	Moderately significant	Very significant	Highly significant	Total
			Effect on efficiency			
Offline	9(36.00)	13(52.00)	2(8.00)		1(4.00)	25
Online	4(26.67)	9(60.00)	2(13.33)			15
Portal				4(36.36)	7(63.64)	11
Total	13(25.49)	22(43.14)	4(7.84)	4(7.84)	8(15.69)	51

Note: Figures in parentheses are row percentages.

transactions. A large number of firms that did not perceive any significant impact of e-business tools were using offline technologies. Whereas all the sample firms that have adopted portal-based e-business tools reported that the impact on efficiency in transactions is very evident. Sixty-four per cent of them found that the impact is highly significant. MDs of more than 50 per cent of the firms said that the impact of e-business technologies is significant. Considering the perception of MDs on this aspect of e-business technologies, we may hypothesize a positive relationship between the adoption of e-business tools and efficiency in business transactions.

4.8.2   Competitiveness (COMP)

Table 2.5 presents the perception of MDs on the impact of e-business technologies on competitiveness of firms.

Table 2.5   Competitiveness of firms

EB-type	Not significant	Significant	Moderately significant	Very significant	Highly significant	Total
		Effect on competitiveness				
Offline	11(44.00)	7(28.00)	7(28.00)			25
Online	3(20.00)	5(33.33)	6(40.00)		1(6.67)	15
Portal	1(9.09)			6(54.55)	4(36.36)	11
Total	15(29.41)	12(23.53)	13(25.49)	6(11.76)	5(9.80)	51

Note: Figures in parentheses are row percentages.

Table 2.5 shows that by and large the pattern of MDs response on the competitiveness of firm *vis-à-vis* the adoption of e-business technologies is similar to that of efficiency in business transactions. A majority (44 per cent) of the offline technology-using firms did not consider a significant impact on the competitiveness of firms while 91 per cent of portal-using firms perceived that e-business technologies significantly strengthen the competitiveness of firm. Hence a positive relationship between the adoption of e-business technologies and the augmentation in the competitiveness is hypothesized.

5   Statistical results

All the variables that are discussed in Section 3 were analysed in a univariate framework. The mean value and standard deviation of these variables are presented in Table 2.6 along with the F-value and the level of significance.

It can be seen from Table 2.6 that the mean value of the variables such as WAGE, STO, EFFITRANS, COMP, and BANDW differs very significantly among the three types of firms. Table 2.6 also shows that the profit margin

Table 2.6 Univariate analysis of variables

	Mean and SD of Variables					
Variable	Type of E-business			F-value	Probability	Remarks
	Offline	Online	Portal			
WAGE	3.596 (2.043)	5.601 (3.269)	11.199 (5.166)	20.564	0.0000	Wage rate in Rs.'000 per month
STO	56.740 (92.933)	125.626 (136.77)	949.682 (1074.34)	13.050	0.0000	Sales turnover in Rs. millions
EXPOINT	0.003 (0.016)	0.139 (0.331)	0.294 (0.419)	4.869	0.0119	Exports/Sales turnover
COL_TECH	0.320 (0.476)	0.667 (0.488)	0.727 (0.467)	3.921	0.0265	Technological collaboration
EFFITRANS	1.840 (0.898)	1.867 (0.640)	4.636 (0.505)	58.169	0.0000	Induces efficiency in business transactions
COMP	1.841 (.851)	2.400 (1.056)	4.091 (1.136)	20.341	0.0000	Strengthen competitiveness
PRM	5.666 (3.348)	4.289 (2.690)	6.095 (4.333)	1.096	0.3424	Profit after tax/ Sales turnover
BANDW	1.080 (0.277)	1.267 (0.458)	2.273 (0.647)	30.021	0.0000	Bandwidth
MDEDU	2.440 (0.917)	2.933 (0.884)	3.455 (0.934)	4.953	0.0111	Education of Managing Director

Note: Figures in parentheses are standard deviations.

does not differ significantly among the sample firms. The other variables such as EXPOINT, COL_TECH, and MDEDU also differ significantly. Subsequently, the data were analysed in a multivariate framework. We have used LIMDEP Version 7.0 econometric software which gives one constant and threshold parameter for index, that is, in a model specification of $z = \beta'x + \varepsilon$

$$y = 0 \text{ if } z <= 0,$$
$$y = 1 \text{ if } 0 < z <= \mu1, \text{ and}$$
$$y = 2 \text{ if } z > \mu1.$$

The results of ordered PROBIT analysis are presented in Table 2.7.

Due to the presence of multicollinearity, several equations were tried. The estimates of four equations are presented in Table 2.7. In first equation the entire model was estimated and all the variables except

Table 2.7 Maximum likelihood estimates of e-business adoption (ordered PROBIT model)

Equations	I	II	III	IV
Constant	−3.374 (−3.986)*	−1.828 (−3.454)*	−2.4661 (−4.387)*	−2.147 (−3.195)*
STO	0.0018 (1.140)			0.003 (2.055)**
EXPOINT	0.452 (0.439)			
BANDW	0.772 (1.803)***		1.746 (4.388)*	
COL_TECH	0.839 (1.740)***	0.892 (2.361)**	0.887 (2.391)**	
WAGE	0.146 (1.776)	0.268 (4.045)*		
MDEDU	0.355 (1.575)			0.633 (2.890)*
Threshold parameter for index ($\mu1$)	1.879 (3.350)*	1.317 (4.034)*	1.388 (4.144)*	1.270 (4.026)*
Log likelihood	−28.185	−36.471	−35.717	−37.530

Note: Significant * at 1%, ** at 5%, and *** at 10%
  Figures in parentheses are z-values.

BANDW and COL_TECH (significant at 10 per cent) turned out to be insignificant. This is due to high correlation among the independent variable (Appendix Table 2.2). We dropped three most insignificant variables from Equation I and re-estimated the Equation II. The estimates of the Equation II show that two variables, namely, COL_TECH, and WAGE have emerged as significant determinants. It can be seen from Appendix Table 2.2 that BANDW is highly correlated with most of the variables. The reason for BANDW being highly correlated with STO could be that the firms that have large size of operation can afford higher bandwidth because of high access costs of communication network in India. Therefore, in Equation III we included BANDW and other variables that were comparatively less correlated with BANDW. The results of equation III show that BANDW has emerged as a very significant factor that influenced the type of e-business technology adopted by the sample firms. In equation IV we included STO and MDEDU that were not highly correlated with each other. The estimates of equation IV reveal that STO and MDEDU play an important role in the adoption of type of e-business technology.

The emergence of STO as a significant variable in influencing the extent of adoption of e-business is in line with the existing literature. Several studies (Pavitt *et al.*, 1987; Siddharthan, 1992) also found a positive relationship between the adoption of new technologies and the size of operation. Firms that have a large size of operation are comparatively better off than other firms where investment in new technologies is concerned. Moreover, large firms are in a better position to appropriate the benefits of new innovations. Sound financial position and appropriability conditions have strong bearings on the adoption of latest technologies and in the investment on innovative activities. Larger firms have the other advantage of employing more skilled workforce, which may be necessary for using new technologies more effectively and efficiently.

The study finds evidence of an important role being played by the telecommunication network in diffusion of e-business technologies. The emergence of bandwidth as an important factor is a case in point. These findings are similar to other studies (Lal, 2001; NASSCOM, 2000). Availability of higher bandwidth is an external factor to firms. It is related to the institutional environment created by the central and local governments. Lal (2001) found that institutional environment including telecommunication and other ICT-related policies have significantly contributed to the diffusion of ICTs in India. NASSCOM (2000) suggests that parity in telecommunication infrastructure will provide cheap and reliable net access. This in turn will lead to faster diffusion of e-business in India.

One may argue that all the sample firms have equal opportunity to access the telecommunication network as all of them are located in the same area. In principle it is true but in practice all the firms cannot afford reliable mode of communication such as ISDN line because of several times higher tariff rates compared to international charges. In this context, institutional infrastructure is unequally accessible to the sample firms. And this is where the role of government comes in. The government can help the entrepreneurs by providing a high speed and reliable telecommunication infrastructure at internationally competitive rates so that it becomes within reach of small firms.

It is found that wage rates significantly differ among the three types of firms. This may be because most of portal-using firms are electronic goods manufacturing firms. These are engaged in manufacturing of electronics components and finished consumer electronics products. These can be treated as high-technology firms compared to the other sector firms such as garments, plastics, and auto components. Because of the skill-intensive nature of the electronics sector, the wages paid by high-tech firms are higher than those paid by the other type of firms. Moreover, most of the portal-using firms are doing business in international markets. In general, export-oriented firms pay higher wages than the firms that are operating in the domestic market. The emergence of wage rate as a significant factor confirms the findings of other studies (Doms *et al.*, 1997; Lal, 1996) on the adoption of ICTs.

The results of the study also suggest that firms that have technological collaboration with foreign firms have adopted advanced e-business technologies. The finding is not contrary to our expectations. Technologically collaborating firms need a greater degree of interaction with the suppliers of technology than other firms. Interaction consists not only in knowing the specification of the imported equipment but also in sharing intangible and embodied knowledge in the technology. That involves an exchange of drawing, detailed design, and regular interaction. With the help of e-business tools, the technology supplier can interact online with the users of technology by audio-visual means. That could be one of the possible reasons for using advanced e-business tools by firms that have technological collaboration with foreign firms.

In this study we have tried to capture the role of qualification of MDs on the adoption of e-business technologies. The results show that the qualification, and knowledge and information base of the entrepreneur played a key role in the adoption of e-business technologies. This is in accordance with the neo-Schumpeterian literature. The role of qualified entrepreneur in implementing Information and Communication Technologies has been given utmost importance by Mansell and Wehn (1998). They concluded that the introduction of ICTs require new forms of organization. These organizational changes need to be identified and implemented by informed managers. The finding of this study is in accordance with the existing literature.

Export-intensity has not emerged as significant factor in influencing the adoption of e-business tools. This may be attributed to the correlation with other factors whose relative importance is higher in influencing the degree of e-business adoption. Similarly, the study does not find any evidence of profit margin playing any role in influencing the adoption of e-business tools. One of the possible reasons could be that monetary gains of e-business are not very high. Advanced e-business tools using firms might have adopted such tools for their survival and to remain competitive in international markets.

6   Summary and conclusion

The chapter identifies and analyses the determinants of the adoption of e-business technologies in the manufacturing sector in India. The data are drawn from 51 firms located in NOIDA. The survey was conducted during December 2000 and February 2001. Firms were revisited during January to March 2005. Sample firms are dominated by SMEs as 73 per cent of these firms employ less than 150 persons. Entrepreneurial characteristics such as Managing Director's education and age, historical data of firms, and other firm-specific factors such as size of operation, export intensity, international orientation, wage rates, profit margins were included in the analysis. We also included a variable, i.e. bandwidth that represents the institutional environment created by central and local governments. The opinions of MDs on potential benefits of the adoption of e-business technologies were also considered.

Sample firms were grouped into three categories, namely, (1) offline, (2) online, and (3) portal-using firms. Firms that were doing e-business using offline technologies such as messaging systems were grouped in the first category, whereas firms that were using messaging systems as well as online e-business tools such as ASPs in their URL were classified as online e-business doing firms. Portal and ERP using firms were considered as advanced users of e-business technologies and were treated in the third category. Firms were assigned ranks depending on the type of e-business technology used by them. As far as the model of e-business is concerned, 92 per cent of firms adopted B2B e-business model.

The multivariate ordered PROBIT technique was used to identify the determinants of the degree of the adoption of e-business tools. The study reveals that the firms that are more internationally oriented have adopted more advanced e-business tools. Wage rates and scale of operations have also emerged as significant determinants of the adoption of e-business technologies. The study captures the role of bandwidth in diffusion of e-business technologies. The findings of the study are by and large similar to other studies (Doms *et al.*, 1997; Bedi, 1999; Siddharthan, 1992; Cohen, 1995; Pavitt *et al.*, 1987; Soete, 1997)

The study shows the evidence of a positive association of type of e-business technology used by firms and the bandwidth. A study by NASSCOM (2000) suggests that availability of higher bandwidth is a prerequisite for the penetration of Internet and web-enabled services in India. This study concludes that a very reliable and affordable telecommunication network has to be in place to harness the potentials of ICTs. The IT Law 2000 is a necessary but not sufficient condition for the success of diffusion of ICTs. The initiatives towards communication technology convergence are expected to result in faster growth of e-business. The findings of the study suggest that there is a need to create proper local, national, and global information infrastructure to derive the maximum benefit from the ICT revolution.

Appendix Table 2.1   Product profile of firms

S. No.	Industrial classification	Products	No. of firms	Per cent
1	Auto-components	Auto Locks, Door Handles, Speakers, Electronic Systems for Cars	5	9.80
2	Electrical Equipments	Electrical Stamping, Special Transformers	2	3.92
3	Electronics	Telecommunication Equipments, Cable TV Hardware, CTVs, CTV Components, Printed Circuited Boards (PCB), PCB Design of Electronic Components, Software Products	17	33.33
4	Garments	Ladies, Children, Socks	5	9.80
5	Machinery	Construction Equipments, Electrical Panels, Industrial Equipments	8	15.69
6	Plastic Products	Plastic Injection and Blow Mould, Components for TV, Plastic Bottles	7	13.73
7	Others	Adhesive, Polypack, Rexene, Vibration-Shock Absorbers, Footwear	7	13.73
				13.73

Appendix Table 2.2   Correlation matrix

	EB-TYPE	STO	EXPOINT	BANDWIDTH	COMP	EFFITRANS	COL_TECH	WAGE	MDEDU
EB-TYPE	1.00								
STO	0.52	1.00							
EXPOINT	0.41	0.29	1.00						
BANDWIDTH	0.69	0.60	0.36	1.00					
COMP	0.65	0.33	0.23	0.54	1.00				
EFFITRANS	0.71	0.42	0.26	0.62	0.64	1.00			
COL_TECH	0.35	0.16	0.21	0.17	0.25	0.18	1.00		
WAGE	0.65	0.47	0.68	0.66	0.44	0.53	0.21	1.00	
MDEDU	0.42	0.31	0.45	0.29	0.34	0.23	0.37	0.55	1.00

3
Globalization, Learning Modes and Opportunities in Malaysian SMEs

1 Introduction

Small and medium-sized enterprises (SMEs) are often regarded as the lifeblood of the economy. The same sentiment is felt in both developed and developing countries. They enable in bringing about innovation, increasing productivity, competition and creating value added activities. As the name suggests, they are highly flexible and are able to explore and experiment new ideas and practices in the quest for efficiency and effectiveness. Lee (2003) argues that SMEs form an integral part of the value system as downstream suppliers or service providers to the larger corporations in the overall production network. Realizing these crucial roles that SMEs play, governments are increasingly aware and have been serious about designing appropriate industrial policies to further harness the potential of SMEs as a catalyst to economic growth of the countries.

Seow (1989) argues that SMEs contribution towards the national economy is in the five main areas. First, SMEs employ higher labour per capital investment. Although SMEs factories do not employ as much labour as the large manufacturing industries; in comparison with the capital invested in the business; they employ more labour per dollar invested. For example, a small manufacturing industry with a paid-up capital of RM^3 250,000 may employ about say 30 full-time workers whereas heavy industries with 10 times this capital invested in heavy machineries may need to employ only ten skilled workers to operate the machine. So SMEs actually help to resolve some of the labour problems in this country.

Second, SMEs assist tremendously towards increasing investment. The principle sources of finance for the SMEs sector are the owners own funds and those which he is able to source from his families, friends and relatives and possibly some close business associates. Most likely a large portion of

---

3 Ringgit Malaysia. Abridged version of the chapter is under consideration for publication in Research Policy.

the capital obtained by the entrepreneurs for investment in small industries would not have been made available to large establishments or to government for investment. They would probably have been used by the entrepreneur in setting up their various businesses.

Third, SMEs play a big role in generating entrepreneurship and creativity. The low cost of setting up a small manufacturing unit enables an enterprising worker not only to provide himself with a livelihood but also to offer employment to others. Therefore, SMEs serve as a training ground for developing the skills of industrial workers and entrepreneurs. The training and experience which the workers obtain in the operation of the SMEs, would ultimately allow a number of them to expand and to branch out on their own. These entrepreneurs who have been trained on the basic skills usually are creative in their own ways and may branch into other related fields and therefore, SMEs generate creativity. This creativity of the entrepreneurs can generate new products, for example wood carving products and pottery.

Fourth, SMEs are also a good training ground for skill development. The establishment of SMEs actually enables the small entrepreneurs to use their skill and knowledge, as well as to improve their technological capabilities. Therefore, this will eventually accelerate the process of technology absorption and dispersion of economic benefits. And finally, SMEs also provide back-up service for the large industries. This is because it may not be cost effective for the large industries to go into component parts manufacturing. It would be more economical if they have their component parts manufactured either by their subsidiaries or sub-contract it to various SMEs which are able to supply them with these components at a cheaper price and sometimes more efficiently.

Many factors currently contribute to the rising importance and relevance of SMEs. First, with the advent of Information and Communication Technologies (ICTs) and improved transportation over the last 50 years, people can travel easily and communicate effectively and efficiently. The direct result of this is people can work together without having to be physically together. Organizations find that they can easily outsource some of their work to smaller companies, thus encouraging the establishment of more service industries. Organizations are also becoming smaller. Second, SMEs are flexible in responding to demand changes. Beaver (2002) describes this consequence as enterprise culture. Entrepreneurship has been long recognized as a major factor of production. Third, SMEs play a substantial role in the value creation and delivery.

Considering the crucial role of SMEs, all stakeholders need to understand the problems and constraints of SMEs and should strive to contribute towards any plan of action for the development of SMEs. Some of the key issues and challenges faced by SMEs in Malaysia during their growth process are as follows.

1.1 Liberalization and global competition

Multilateral and regional trade and investment liberalization policies have made markets more accessible and competition more intense among local manufacturers. It is becoming more important to be internationally competitive in order to function effectively even in the domestic market. In today's fast changing and dynamic environment, it is essential to achieve and maintain one's competitive edge. SMEs in this country lack the proper motivation, assistance in acquiring necessary skills, strong infrastructure support and the provision of right advice and assistance.

1.2 New emerging technologies

Most SMEs in this country do not see the significance of ICT applications in their daily operations. They are not aware that emerging technologies and advances in ICTs have contributed to productivity growth and competitiveness. The increasing ICT usage in business makes it possible for large corporations to seek multiple suppliers in their value chain. These suppliers have to meet product quality, cost and speedy delivery requirements. As leading edge technologies increasingly become the determinants of competitiveness, SMEs in this country need to develop their capacity to adopt and adapt technologies that are appropriate and relevant to their industries.

1.3 Skills development

Most SMEs in Malaysia are labour-intensive and are continuously dependent upon foreign workers. They are also very slow in investing in automation, skills upgrading and knowledge acquisition, which are all important to long-term competitiveness. Generally, SMEs in this country have a negative attitude towards training investment. They fear losing well-trained staff to competitors together with their investment. Therefore, the lack of supply of skilled and knowledgeable workforce stifles output expansion. As such, there is a strong need to instil a training culture among SMEs in order to produce skilled and knowledgeable workers for SMEs to be competitive in terms of quality, price, and delivery.

1.4 Finance

The most prevalent problem faced by SMEs in Malaysia is the inadequacy of finance. The main reason is that SMEs are perceived as high-risks by the average banker and therefore, bankers are reluctant to approve lending. SMEs traditionally finance their operations through supplier credits, own savings, loans from family and friends. The SMEs which are newly set-up face difficulties in obtaining credit, as they have little collateral and no sound track record. A longer duration is needed for their loan processing due to their high business risk. Therefore, there is an urgent need to improve access to institutional credit for SMEs in this country because a loan delayed is virtually a loan denied.

1.5 Information

The ability to seek and apply information in business operations will definitely assist SMEs in terms of efficiency in the new business environment. SMEs need to acquire important and critical knowledge and skills to remain competitive. Technology and knowledge investments have provided significant competitive edge to companies especially in terms of design, product research, process innovation and management information systems. Therefore, SMEs need a reference centre where they can access information and advice on various areas concerning their operations.

In view of the globalization process which is reshaping national economies and the importance of SMEs, this study seeks to identify the role of institutional environment in the diffusion of ICT led new technologies which is expected to strengthen the competitiveness of firms. Although ICT led technologies are regarded as path independent, it is worthwhile to understand the growth pattern of SMEs and the changes in their structure for the last two decades. The main objectives of the study are:

- To provide an overview of SMEs in Malaysia
- To identify and analyse the policies of institutions that are meant for promotion of SMEs
- To examine the technological profile of sample firms
- To identify and analyse factors that influence the adoption of ICTs

The organization of the chapter is as follows: Section 2 presents the historical background of SMEs. In Section 3 we present the definitions of SMEs, and the growth and economic contributions of SMEs to the national economy. Section 4 describes the institutional support to make SMEs globally competitive while Section 5 underscores the methodology and survey of firms. Section 6 describes characteristics of sample firms. Entrepreneurship and the degree of the adoption of ICTs are discussed in Section 7 while Section 8 makes available the statistical results. Findings are summarized in Section 9.

2 Historical background of SMEs

Historically, names used to describe the manufacturing industries of South East Asia include not being of large-scale or capital-intensive organizations (Boeke, 1971). Boeke (1971) termed them household, handicraft, cottage, workshop and small factory-like businesses; all possessed little capital and only the small factory-like used electrical power. According to Boeke (1971), cottage and small industries were prevalent in South East Asia before the western capitalistic influence, and few have survived largely unchanged until today. However, the majority were unable to withstand the competition of imported factory-made manufactures. Fryer (1978) concludes that most Asian countries emphasized the promotion of cottage and small-scale industries in their industrialization plans. In Third World countries like Malaysia, Indonesia and Thailand the industrialization drive was a crucial part of their nationalist movement.

However, in Malaya (former name of Malaysia before independence in 1957), milder nationalism and much higher per capita income as compared to other Third World countries during the period before 1957; had permitted a policy of allowing private and foreign capital to assume greater part of their new industrial development. In the case of Malaya, it had encouraged cottage and small industries on a co-operative basis.

In addition, traditional cottage industries in Malaysia were concentrated in areas where alternative forms of employment were very low and the cost of labour was cheapest, such as in the handloom industry of Kelantan and Terengganu. According to Fisk (1962), 95 per cent of Terengganu weavers during the period in which the research was done, were illiterate in any languages. He also mentioned that payments to weavers constituted one-third of the gross product of the East Coast handloom industry in Malaya and the average worker's salary during that time was less than 16 cents per hour. Fryer (1978) also concluded that many cottage and small-scale industries in Malaya had high import content. About half of the payments made by the Malayan handloom industry were in respect of imported raw materials. He also claimed that imported materials were used because local raw materials often were unstable in terms of supply and quality.

In Malaya, smokehouses for smallholder rubber constituted by far the most important cottage industry. They were located in close proximity to the houses of their operators and they relied almost entirely upon the labour of the operators and their families. The smokehouses usually acted as service centres. They retained the ownership of their sheets after smoking and the sheets were then sold to the dealers. Sometimes, the smokehouse owner might also be a dealer and might buy smoked or 'unsmoked' sheets. Smokehouse operators were mostly dominated by the Chinese community that provided valuable service for the smallholders at a modest fee which central processing stations producing in bulk could not match. They were mostly located in areas of smallholder production as latex was seldom transported for more than 6 miles by smallholders and the collecting cost was negligible.

As for the small-scale industries which used electrical power, Malaya was lucky that its per capita electrical power generation was comparatively higher than other South-East Asian countries. However, an interconnected network only existed in the west central portion of the country and there were many gaps within the system as well. Malaya's proportionately higher urban population also contributed to industrialization as electricity was cheaper when it was supplied to various classes of users, ensuring a higher load factor. All these helped foster the development of the small-scale industries in Malaya during that time. In the 70s, 80s and 90s, the Malaysian Government had encouraged large-scale industries to provide more markets for SMEs by granting large enterprises in this country 'pioneer' status and the remission of taxes, duty free imports of raw materials and granting of sole production rights for a period of time. All these were undertaken to achieve the above mentioned purpose.

3 Definition of SMEs and economic contributions to national economy

This section is divided into three sub-sections. In the first sub-section we discuss the definitions of SMEs that are used by various institutions in Malaysia and elsewhere. An overview of SMEs is presented in the next sub-section while in the last sub-section economic contributions are examined.

3.1 Definitions of SMEs

Beaver (2002) has highlighted that it is easier to describe than to define SMEs. One popular definition adopted by the Bolton Committee in UK is: (a) In economic terms, a small firm is one that has a relatively small market share, (b) it is managed by its owner in a personalized way and not through the medium of a formalized management structure, and (c) it is independent from outside control and interference in taking their principal decisions. In addition, the Committee also specifies different threshold number of employees for different sectors. Manufacturing is limited to 200 employees whereas for retailing, a turnover of £ 50,000 is used as a criterion. European Commission has adopted a common definition of SMEs with more emphasis on numbers of employees instead. They define micro-enterprises (very small firms) as employing less than ten workers, small enterprises as those with ten to 99 workers, and medium enterprises as those with 100 to 499 workers.

Malaysia is no exception and there is no standard definition of SMEs. According to Lee (2003), different government agencies have adopted different definitions based on criteria such as annual sales turnover, number of full-time employees and shareholders' funds. International Cooperative Alliance defines SMEs as enterprises with full-time employees of less than 75 and shareholder funds of less than RM 2.5 million. The Central Bank of Malaysia (known locally as Bank Negara) defines SMEs as enterprises with shareholders' funds of less than RM 10 million. One commonly used definition is provided by the Ministry of International Trade and Industry as enterprises with annual sales turnover not exceeding RM 25 million and full-time employees not exceeding 150. According to Small and Medium Industries (SMI) Association of Malaysia, SMI companies should be incorporated under Malaysian Companies Act 1965 must have at least 70 per cent of their shareholdings held by Malaysians and are involved in the manufacturing sector. An SMI company is also a company with full-time employees not exceeding 150 and with an annual sales turnover not exceeding RM 25 million. Their classification is as follows: (a) small-scale Company is a company with an annual sales turnover of less than RM 10 million and not more than 50 full-time employees, and (b) medium-scale Company is a company with an annual sales turnover between RM 10 million to RM 25 million and with 51 to 150 full-time employees. However, the definitions provided by SMI Association of Malaysia only refer to companies in the manufacturing sector.

Other definitions provided by Small and Medium Industries Development Corporation (SMIDEC) are more appropriate and comprehensive given the importance of the service sector and the robust development of ICT companies in Malaysia. SMIDEC classifies SMEs into two broad categories:

a. Manufacturing, manufacturing-related services and agro-based industries. The characteristics include:
- Micro-enterprise has sales turnover of less than RM 250,000 or full-time employees less than five.
- Small enterprise has sales turnover between RM 250,000 and less than RM 10 million or full-time employees between five and 50.
- Medium enterprise has sales turnover between RM 10 million and RM 25 million or full-time employees between 51 and 150.

b. Services, primary agriculture and information and communication technology. The characteristics include:
- Micro-enterprise has sales turnover of less than RM 200,000 or full-time employees less than five.
- Small enterprise has sales turnover between RM 200,000 and less than RM 1 million or full-time employees between five and 19.
- Medium enterprise has sales between RM 1 million and RM 5 million or full-time employees between 20 and 50.

3.2 An overview of SMEs in Malaysia

In 2002, SMEs formed nearly 96 per cent of the total 212,888 companies registered in Malaysia (see Table 3.1) and contributed 29.1 per cent of the total manufacturing output, 26.1 per cent of value-added. This means that in 2003, SMEs provided about a third of the total employment (32.5 per cent) in the country (MITI Statistics, 2003). In 2004, about 40 per cent of the SMEs in the manufacturing sector are in the resource-based sector. Several important sectors in the SMEs include the wood and wood products industry (12.8 per cent) food, beverages and tobacco (10 per cent), paper and paper products (9.9 per cent), and rubber and plastic industry (7.4 per cent), machinery and equipment (9.4 per cent) and textiles, apparel and leather (8.8 per cent).

Table 3.1   Distribution of firms by size (2000)

Type	Manufacturing	Service	Total	Share (%)
Micro	7,171	114,840	122,011	57.31
Small	9,445	53,612	63,057	29.62
Medium	1,655	17,976	19,631	9.22
Total SMEs	**18,271**	**186,428**	**204,699**	**96.15**
Large	2,090	6,099	8,189	3.85
Total	**20,455**	**192,527**	**212,888**	**100**

Source: Government of Malaysia, Department of Statistics, Census 2000.

Table 3.2 presents the sectoral distribution of manufacturing SMEs in 2002. The largest number of establishments of SMEs was the textile and apparel sector (18.2 per cent), followed by food and beverages (15.2 per cent), metal and metal products (14.8 per cent) and wood and wood products (14.1 per cent). Percentage of SMEs in most of the sectors was higher than 90 per cent except in plastics, electrical and electronics, petrochemical, rubber, and palm oil sectors. The very low percentage of SMEs in

Table 3.2   Distribution of SMEs in the manufacturing sector (2000)

Sector	Number of total establishments	SMEs	% of SMEs	Sectoral composition of SMEs (%)
Textiles & apparel	3,419	3,319	97.08	18.2
Food & beverages	2,949	2,749	93.22	15.2
Metal & metal products	2,918	2,709	92.84	14.8
Wood & wood products	2,776	2,582	93.01	14.1
Paper, printing, publishing	1,288	1,195	92.78	6.5
Machinery & engineering	1,249	1,135	90.87	6.2
Plastics products	1,121	988	88.14	5.4
Electrical & electronics	907	543	59.87	3.0
Non metallic mineral products	893	803	89.92	4.4
Others (Jewellery)	733	666	90.86	3.6
Petro-chemical & chemical	712	526	73.88	2.9
Transport equipment	507	433	85.40	2.4
Rubber & rubber products	482	366	75.93	2.0
Palm oil & palm oil products	434	155	35.71	0.8
Leather	67	65	97.01	0.4
Total	**20,455**	**18,271**	**89.32**	**100.0**

Source: Department of Statistics, Census 2000.

electrical & electronic, compared to other sectors, (59.87 per cent) may be attributed to capital and skill intensive nature of the sector.

3.3   Economic performance of SMEs

In general, SMEs play a crucial role in the Malaysian economy because they act as the backbone of industrial development and widening of the industrial base in this country. Tables 3.3 and 3.4 present the economic performance of SMEs during 2002 and 2003. We are unable to present long-time series data due to inadequate reliable sources.

Higher growth rate of value added than employment suggest that value added per employee has increased from 2002–2003. Either firms have been able to augment labour productivity through the adoption of new technologies or existing SMEs have altered product profile and reoriented themselves toward the products where higher value added has been realized. Another reason could be that new SMEs might have been setup in high-

Table 3.3   Output, value added and employment in SMEs

Indicators	Value	% Share to the manufacturing		Annual growth rate	
	2003	2002	2003	2002	2003
Total output (RM billion)	68.9	62.8	29.1	29.1	9.7
Value added (RM billion)	14.2	12.7	26.1	25.8	11.8
Employment (Number)	375,840	362,345	32.5	31.6	3.7

Source: National Productivity Corporation (2003).

Table 3.4   SMEs productivity indicators (2003)

Productivity	Value	Annual growth rate (%)
Labour productivity		
Output per employee (RM)	183,222	5.6
Added value per employee (RM)	37,675	7.8
Labour cost per employee (RM)	18,762	7.3
Added value per labour cost (pure number)	2.0	0.4
Unit labour cost (pure number)	0.1	1.6
Capital productivity		
Fixed assets per employee (RM)	35,792	4.0
Added value per fixed asset	1.1	3.6

Source: National Productivity Corporation (2003).

tech sectors such as consumer electronics and information technology and networking products where higher value addition can be relatively easily realized than traditional sectors such as food and textiles. Table 3.4 presents few more indicators of economic performance of SMEs.

4   Institutional support to SMEs

Two main policy measures have been initiated to make SMEs globally competitive. First set of policies aimed at helping SMEs financially while the other set of initiatives focus on human resource development and technology acquisition policies. First we discuss the institutions and their financial policy and subsequently we will present technology-related policy measures.

4.1   New initiatives to improve access to financing for SMEs

The National Small & Medium Enterprise Development Council has recently agreed to the introduction of an interest subsidy and securitization of SMEs' loans to encourage further lending by financial institution to SMEs. The interest subsidy provided by the Government will enable viable SMEs with higher risk profile to obtain financing at lower costs. The ability of financial institutions to securitize SMEs' loans in their portfolio will further increase their capacity to lend to SMEs.

The access to financing by SMEs has increased substantially, as evidenced by the high rate of utilization of special funds made available by Bank Negara Malaysia to SMEs. During the first 9 months of 2004, a total of 3,380 SMEs obtained new loans totalling RM 2.2 billion under the special funds. In 2004, Bank Negara Malaysia raised the total allocation for the existing five special funds by RM 3.4 billion to RM 8.9 billion due to the high demand for loans from SMEs. Of this amount, RM 7.4 billion or 82 per cent was approved to 17,928 SMEs as at the end of September, 2004.

Export Credit Refinancing (ECR) is also another form of financing incentive for SMEs. It provides short-term credit at preferential interest rates to Malaysian exporters. The facility is to promote the exports of manufactured goods. ECR is operated by commercial banks, which are refinanced by the Export-Import Bank of Malaysia (EXIM Bank) for the export credit extended to eligible exporters. The exporters may invoice his export in any currency but financing is only in Malaysian Ringgit. Certain conditions are applied for granting loans under this scheme.

The transition to a knowledge-based economy has also been expedited with the provision of RM 500 million in venture capital funding in the 2001 Budget. A portion of the RM 500 million funds was dedicated to SMEs start-ups and high-growth concerns. This is to ensure that adequate seed capital is channelled to the manufacturing sector. The Malaysian Venture Capital Management Bhd was set up under the Ministry of Finance as a one-stop centre that interfaces with SMI development agencies to ensure greater coordination in the implementation of venture capital financing.

4.1.1   Funding arrangement to Participating Financial Institutions (PFI) for SMEs financing

The funding for credit facilities under special funds for SMEs financing is provided by the Government. This is done by placing the same amount of fund with the PFI for on-lend to the SMI. The Government would impose a funding rate with the PFI, which is normally much lower than the market rate. The PFI would, in turn, lend the credit facility to SMI at a certain margin to cover the funding cost, administrative expenses as well as the credit risk. The interest rate is to be determined by the Government. Due to this one-of-a-kind funding arrangement, the interest on lending to SMI under the various special funds would normally be at a concessionary rate. Figure 3.1 shows the work flow of a government-funded loan scheme.

Six steps are required to get loan under this scheme. First, SMI submits loan application to PFI. Second, applications are scrutinized by PFI and then PFI requests for fund from Government. Third, step government places funds with PFI. Fourth, loan is then disbursed to SMI by PFI. Fifth, SMI repays loan to PFI. And finally in sixth step the funds are returned to the government by PFI.

4.2   SMI-related special funds

In 1988, the Government established the first fund namely the New Investment Fund and since then, many funds have been established through the various ministries and agencies. To date, there are more than 40 funds available for the SMI to finance their various business fields. The respective ministries and agencies act as 'Fund Administrators' to ensure that the implementation and utilization of the funds conform to the guide-

Figure 3.1   Work flow of a government-funded loan scheme

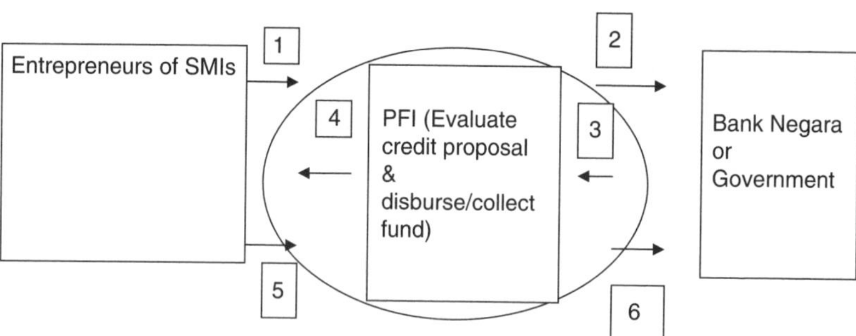

Source: SMIDEC (2002).

Table 3.5   SMI-related special funds

No.	Name of fund	Objectives	Fund administrator
1	Fund For Small & Medium Industries 2 (Conventional & Islamic)	To promote SMI activities in both the export & domestic-oriented sectors.	Bank Negara Malaysia
2.	Rehabilitation Fund For SMI (Conventional & Islamic)	To assist viable SMIs facing temporary cash flow problems.	Bank Negara Malaysia
3	Fund For Food (Conventional & Islamic)	To increase food production in Malaysia.	Bank Negara Malaysia
4	New Entrepreneurs Fund 2 (Conventional & Islamic)	To help stimulate the growth of SMI Bumiputra enterprises.	Bank Negara Malaysia
5	Entrepreneur Rehabilitation & Development Fund (Conventional & Islamic)	To assist distressed but viable Bumiputra companies through non-performing loans/ non-performing financing (NPL/NPF) resolution and additional working capital support.	ERF Sdn Bhd

Source: SMIDEC (2002).

lines and Government objectives. Some of the popular funds presently related to SMI are listed in Table 3.5.

4.2.1   *Fund for Small & Medium Industries 2 (FSMI2)*

In October 1997, the Minister of Finance set up the Fund for SMI (FSMI) to strengthen linkages between SMI and the larger industries. The objective is also to facilitate SMI in penetrating the export market. The Government had allocated an initial amount of RM 1 billion fund for the SMI in January 1988 and the fund was subsequently increased to RM 1.5 billion in May 9 1998.

The purpose of this fund is to promote new investments in the selected sectors such as manufacturing, agro-based and special service industries as part of the Government's efforts to promote exports and new growth industries including high technology industries. The number of FSMI loans approved has increased sharply over the years. Following the termination of FSMI in April 2000, the Fund for FSMI2 was introduced with an allocation of RM 200 million. The Fund was doubled to RM 400 million in March 2001. The allocation as at March 26, 2003 totalled RM 1.25 billion (Lee, 2003).

4.2.2   Rehabilitation Fund for SMI (RFSMI)

Whilst the FSMI2 is a relatively cheaper source of financing for SMI, the fund was extended primarily to finance working capital requirements. Most of the SMI are currently facing non-performing loan problems. In view of this, the Government has set up the Rehabilitation Fund for SMI whereby part of the credit facilities could be utilized to restructure the non-performing loans (NPL) of SMI. The Government had in November 19, 1998 announced an allocation of RM 750 million for the Rehabilitation Fund for SMI (RFSMI). The main objective of establishing this fund is to provide financial assistance to viable SMI (with at least 75 per cent of the amount approved to be channelled to Bumiputra (native) SMI) which are facing NPL due to temporary cash flow problems.

Similar to FSMI, the purposes of credit facility can be used for expansion of productive capacity or working capital or both. In contrast to FSMI, the credit facility of up to 40 per cent of the amount approved can also be used to refinance or restructure existing NPL or facilities with an existing PFI. The remaining 60 per cent of the fund must be used to expand productive capacity or as working capital. RFSMI was designed to expedite the resolution of NPL problems of SMI for viable projects in the manufacturing, agro-based industry of services sectors.

The lending under RFSMI is encouraging but growth is at a slower rate. Moreover, to ensure an effective utilization of the Government's financial resources, the total allocation for RSFMI was reduced from RM 750 to RM 500 million with effect from August 4, 1999 and the excess allocation of RM 250 million was transferred to FSMI to meet the high demand for financing under FSMI. The lending under RFSMI in terms of number of accounts and amount outstanding as at December 31 2002 totalled 297 and RM 322.6 million respectively.

4.2.3   Fund for Food (3F)

The Fund for Food with an initial allocation of RM 300 million was launched in January, 1993 with the main objective of promoting investment in food sector, particularly, for primary food production. The setting up of Fund For Food or commonly known as 3F Scheme was part of the supply policy response to complement monetary measures to reduce inflation and in particular, the rising prices of food and beverages.

Due to the encouraging response, the allocation was increased to RM 700 million in 1997 and in May 1999 the allocation was further increased by another RM 300 million to RM 1 billion. It was prompted not only by the good response to the 3F Scheme but also the need to refocus resources to the agricultural sector in order to complement the Government's efforts to reduce Malaysia's food import bills. An additional allocation of RM 300 million was approved by the Minister of Finance in 2001.

The 3F Scheme offers financing to Malaysian citizens and as well as Malaysian-controlled companies that have at least 51 per cent shareholdings. The target groups for this scheme are the larger and medium-sized food projects. However, it is a requirement that the investment must be in Malaysia and are domestic-oriented projects, where at least half of the total production is sold in the domestic market.

4.2.4   New Entrepreneurs Fund 2 (NEF2)

As part of the New Economic Policy, the New Entrepreneurs Fund (NEF) was launched on December 12 1989 with a total allocation of RM 250 million. The primary objective of the NEF is to assist the Bumiputra entrepreneurs to venture into the various fields of businesses particularly in the productive sectors of the economy that include manufacturing, agriculture, tourism and export-oriented businesses. Due to the overwhelming response to the Fund, the size of the Fund has been raised in stages since 1989 to the current allocation of RM 1.25 billion. NEF2, with a new allocation of RM 500 million was formed following the termination of NEF effective July 15, 2001. The allocation of NEF2 was increased to RM 900 million from March 26, 2003.

Due to the tight liquidity situation, the maximum lending rate under the NEF was raised from 5 per cent per annum to 8 per cent per annum in July 1998. Consequently, the maximum lending rate under the Fund was reduced to 6 per cent per annum with effect from May 3, 1999 following the easing of monetary policy and the lower interest rate environment. The current interest rate is fixed at 5 per cent per annum. The maximum credit facility available in this fund is RM 2 million over a maximum tenure of eight years. The application procedure is similar to FSMI2 and RFSMI, which is through any of the PFIs for credit facilities under the Fund. Since its conception, disbursements of the Fund have been channelled largely to the manufacturing, trading and services sectors, which are areas where Bumiputra participation is still relatively low. In this respect, the Fund has contributed significantly to the general well-being and development of Bumiputra entrepreneurs in these areas.

All small and medium-sized Bumiputra enterprises participating under vendor development programmes of the Ministry of Entrepreneur Development or Ministry of Finance are also encouraged to apply under the Scheme.

4.2.5   Entrepreneurs Rehabilitation & Development Fund (ERDF)

In July 2001 the Entrepreneurs Rehabilitation & Development Fund was established by Bank Negara Malaysia with an allocation of RM 500 million to assist Bumiputra Entrepreneurs to address the problems of NPL as well as to provide working capital. As at December 31, 2002, only 21 loans with a total amount of RM 1.9 million were approved under the scheme.

The aim of ERDF is to assist viable small and medium Bumiputra-controlled companies to resolve the NPL status with one or more financial institutions. Of this allocation, RM 120 million has been set aside for the resolution of all SMI with NPL not exceeding the total limit of RM 1 million. The remainder is needed to assist other businesses with their expansion plans through the provision of additional working capital. The provision of funding support under ERDF is unique. The assistance extended to the eligible borrower not only provides a new working capital but also complements a grant for restructuring the existing NPL through an appropriate debt reduction programme. Summary of funds made available to SMIs in 2001 and 2002 are shown in Table 3.6.

4.3   Bumiputra Entrepreneur Project Fund

Besides ERDF, Entrepreneurs Rehabilitation Fund (ERF) also grant loans on its own under Bumiputra Entrepreneur Project Fund. This fund is allotted for Bumiputra entrepreneurs who have been awarded contracts or projects by the Government or Government-related agencies, statutory bodies and reputable private or public companies. The maximum amount of loan is up to 60 per cent of the contract value less advance given by the Government (if any) or RM 3 million whichever is lower. The interest is fixed at 5 per cent per annum for a maximum period of 5 years. Collateral is not necessary. Bumiputra entrepreneurs or enterprises which are wholly-owned by Bumiputras with viable contracts in hand are eligible to apply for this facility. The other conditions that are applied while granting loans are mainly related to the entrepreneur's eligibility of being Bumiputra. If an application is approved, the release of progressive funds for the project will be closely monitored and controlled by the staff of ERF Sdn Bhd.

4.4   Co-ordination of training & human resource development for SMEs

The Human Resource Development Bhd (HRDB), an agency under the Ministry of Human Resources, has been appointed as the coordinating authority to oversee and coordinate overall training and human resource development for SMEs. Therefore, the role of HRDB will be expanded. This includes the following functions: (a) identification of training needs of SMEs across all sectors of the economy, (b) coordination, monitoring and evaluation of SMEs training and development programmes conducted by Ministries and Agencies, and (c) development of a formal training accreditation system for SMEs.

4.5   Enhanced management and publication of SMEs information

The role of the Secretariat function, which is currently under Bank Negara Malaysia, has been expanded to oversee and coordinate the monitoring, evaluation and publication of SMEs statistics and reports. Apart from

Table 3.6 Summary of special funds from Bank Negara Malaysia

No.	Type of Fund – Date established	Objective	Allocation (Minimum & maximum lending amount)	Lending rate (Tenure)	Total as at end of 2001 & 2002			
					No. of approval		Amount approved (RM million)	
					2001	2002	2001	2002
1	Fund for SMI 2 (FSMI2) 15.04.00	To promote SMI activities in both export & domestic-oriented sectors	RM 400 million (1) Min : RM 50,000 Max : RM 3 million	5% (3 years)	752	1929	357.3	1,048.9
2	Rehabilitation Fund for SMIs (RFSMI)	To provide financial assistance to viable SMIs which have NPLs and temporary cash flow problems	RM 500 million Min : RM 50,000 Max : RM 5 million	4.5% (7 years)	271	297	280.1	322.6
3	Fund for Food (3F) 04.01.93	Promote investment in food sector particularly for primary food production	RM 1.3 billion Min : RM 10,000 Max : RM 3 million (Specific circumstances can be RM 5 million)	3.75% (8 years)	5,201	6,946	1,024.9	1,295.7
4	New Entrepreneurs Fund 2 (NEF2) 15.07.01	To assist the Bumiputra entrepreneurs to venture into the various fields of business	RM 500 million (2) Min : RM 10,000 Max : RM 2 million	5% (8 years)	356	1,336	176.2	652.2
5	Entrepreneurs Rehabilitation & Development Fund (ERDF) 03.07.01	To provide financial assistance to ailing Bumiputra enterprises with viable on-going projects and good recovery prospects.	RM 500 million Min: – Max: RM 3 million	5% (5 years)	–	21	–	1.9

Source: SMIDEC (2002).

meeting the information requirements of SMEs policymakers, the Secretariat has been made responsible for developing Key Perfonance Indicators (KPI) to monitor and evaluate the progress of SMEs development, and will oversee the management of the National SMEs Database which is being set-up by the Department of Statistics. The Secretariat also publishes annual reports on SMEs development for the benefit of policymakers and the SMEs, and thereby improving the availability of SMEs information in the market.

The SMEs Information & Advisory Centre has been set up to exploit the power of ICTs in networking, delivery of support services and web content to SMEs while building up a comprehensive database on SMEs. The Centre will be used for a virtual business matching and for collective purchasing of services and content for SMEs. It is expected to transmit up-to-date and real-time information to SMEs. The Centre reinforces the technical advisory programme which is a key feature in the delivery of SMEs assistance.

4.6 Head Start 500 programme

The Head Start 500 programme was designed by the Government to provide SMEs with support services needed to accelerate their transformation from domestic-oriented to global manufacturers. The 500 enterprises chosen to participate in this programme will provide a tangible measure of success in achieving the targets of the SMI Development Plan. As a start, the programme focuses on SMEs participating in the Industrial Linkage Programme (ILP) and Global Supplier Programme (GSP), the Annual Enterprise 50 Awards, National Productivity and Quality Awards and the Industry Excellence Awards.

4.7 Strengthening technical advisory services to SMEs

Advisory services that enable on-site nurturing of SMEs were implemented to diffuse knowledge, information and skills among SMEs. This complements the business advisory services undertaken through the SMEs Information & Advisory Centre. The programme entails the emplacement of industry experts directly in SMEs to transfer know-how and technology. In addition, on-site assistance is provided by experts with industrial experience.

4.8 Tax incentives

Effective from Year Assessment 2003, small- and medium-scale companies with paid-up capital of RM 2.5 million and below are eligible for: (a) a reduced corporate tax of 20 per cent on chargeable income of up to RM 100,000, the corporate tax rate on remaining chargeable income is maintained at 28 per cent, and (b) dividends distributed is given a tax credit of 20 per cent in the hands of the shareholders.

4.9   Relocation incentives for SMEs

To assist SMEs to relocate to designated industrial sites. The Government provided relocation incentives that directly benefit SMEs. This measure enables SMEs to acquire industrial properties which they can later pledge to banks as collateral for loans.

4.10   Formation of Small & Medium Industries Development Corporation (SMIDEC)

The Small & Medium Industries Development Corporation (SMIDEC) was established on May 2, 1996 to further promote the development of SMI through the provision of advisory services, fiscal and financial assistance, infrastructural facilities, technologies enhancement, human resource development, market access and other support programmes. SMIDEC strives to create resilient and efficient SMEs able to compete in the liberalized market. The Corporation is expected to promote SMEs to be an integral part of the national industrialization programme and in becoming global and world-class manufacturers.

5   Methodology and sample survey

Questionnaires were distributed and personally administered to 100 respondents drawn from different industries. In all there were 67 usable returns, yielding an overall response rate of 67 per cent. The survey was conducted from October 2004 to March 2005. Data were collected through a semi-structured questionnaire provided by UNU-MERIT. Questionnaires were distributed to 100 organizations personally to ensure response from only senior managers such as Chief Operating Officers (COO), General Managers or anyone who held high positions. Finally interviews were also conducted to clarify information collected where and when necessary to ascertain that data and information collected are accurate and reliable.

5.1   Qualitative research approaches

Qualitative research was used to augment quantitative research work. The main aim was to have a better understanding of the network of actors and their policies aimed at promoting SMEs. In this matter, qualitative research approaches proposed by Unesco/Aidcom were widely adopted.

5.1.1   Focus Group Discussion (FGD)

In FGD, the researchers gathered a small group of experts. Researchers also acted as facilitators and guided the interview. FGD was used in this research because it provided an opportunity to capture, share and create new knowledge among the participants who have vast experience in their respective fields. The main aim was to gain a better understanding of a particular phenomenon and how these experts think about a particular issue.

5.1.2   In-depth interview

Researchers conducted detailed interviews with the interviewees who did not want to participate in FGD due to time or distance constraints. However, this was only limited to a small sample. This was carried out mainly through email and telephone interviews with respondents from other states.

5.1.3   Key informant interview

In key informant interview, the interviewees give information about others or specific situations or conditions existing in the area. The key informants are experts who are in a position to provide invaluable insights on some educational issues. Key informants in this research were SMEs owners, SMEs employees, SMEs Association members and government officials involved in SMEs policies.

5.1.4   Content analysis

Content analysis gained popularity a century ago and is used in many fields. UNESCO/AIDCOM (2001) defines content analysis as a technique for gathering and analysing the content of text. The content refers to anything that can be communicated such as words, meanings, pictures, symbols or ideas. Text, on the other hand is anything written, visual or spoken that serves as a medium for communication. In this research, text includes books, newspaper, magazine articles, advertisements and documents. A comparison of strengths and weaknesses of various approaches is presented in Table 3.7.

6   Sample firms' characteristics

Sample firms belonged to several sectors. The largest percentage (35.82 per cent) of firms came from hardware and machinery sector followed by chemical and pharmaceuticals accounting for 14.93 per cent of sample firms. Wood industry emerged as the third largest sector contributing 11.94 per cent of firms. The other sectors that are represented by sample firms are: electrical and electronic, fashion and textiles, and food and beverages. Types of new technologies considered in this study are by and large led by ICTs. Broadly, they can be classified into production and non-production processes. Non-production processes include marketing, coordination of business activities with local and foreign partners and within firms. Such technologies considered in this study are: email, Internet, web enabled technologies, Portal, Management Information System (MIS). Production technologies[4] that are included in this study are:

---

4   Computer Aided Design/Computer Aided Manufacturing (CAD/CAM); Computer Aided Engineering (CAE)' Flexible Manufacturing System (FMS); Numerically Controlled Machine Tools (NCMT).

Table 3.7   Summary table – strengths and weaknesses of various approaches

Research approaches	Strengths	Weaknesses
Focus Group Discussion	• It allows for the collection of preliminary information about the topic. • This approach is comparatively quick and cost effective • There is flexibility in question design and follow up. • The responses are more complete and less inhibited than those from individual interviews.	• The group may become dominated by individuals. • Gathering quantitative data is inappropriate. • A lot depends on moderator's skill.
In-depth Interview	• A wealth of detail is provided. • More accurate responses are provided on sensitive issues. • There are certain situations where interviewees do not wish to be with others.	• Generalization is sometimes a problem. • There are problems in data analysis. • It is time consuming.
Key Informant Interview	• It generates data, which give a broader perspective of the community and provide base-line information that is necessary for the in-depth interviews and the follow up research activities • Key informants are more easily identifiable and their responses are less likely to be biased.	• Key informants do not generate quantitative data. • Findings are susceptible to biases.
Content Analysis	• A researcher can compare content across many texts and analyse with quantitative techniques. • There are qualitative versions of content analysis can be obtained.	• There may be too much data to analysed.

Source: UNESCO/AIDCOM (2001).

CAD/CAM, CAE, FMS, and NCMT. The intensity of the adoption of these technologies by sample firms is presented in Figure 3.2.

All the firms except one used email and the Internet. The firm that did not use either email or the Internet was ice making firm. Given the nature of the product profile of firm, new technologies were not very relevant for the firm. Ninety-seven per cent of firms had their presence in cyberspace. A fairly large percentage (74.6 per cent) of firms had portals for on-line business application processing. Most of these firms used shared portals.

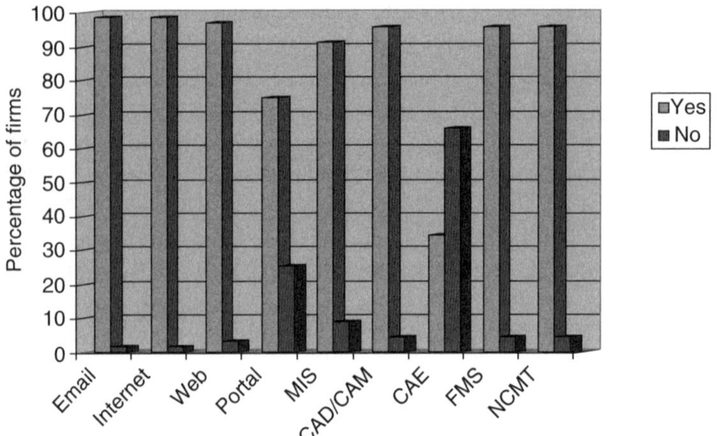

Figure 3.2   Intensity of ICT adoption

Ninety-one per cent of sample firms used MIS but it was limited to in-house management applications. More than 95 per cent of firms used new production technologies except CAE. Merely 34.3 per cent of the firms had adopted CAE. Low usage of CAE could be attributed to its applicability to the sectors to which they belong to. It may be mentioned that CAE is more useful in engineering application.

It was found during the survey that incidence of foreign collaboration of sample firms was very high. A fairly large number of firms (88.1 per cent) had foreign technological collaboration. Nature of foreign technology transfer varied from import of machinery, training, and import of drawing

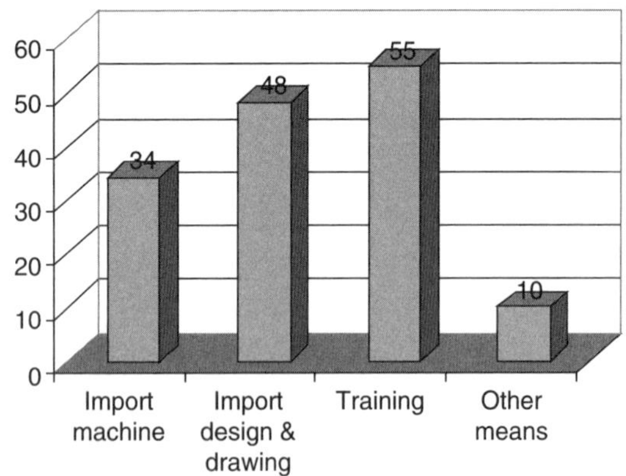

Figure 3.3   Mode of technology transfer (foreign collaboration)

and design. Modes of technology transfer prevalent in firms are depicted in Figure 3.3.

The largest number of firms (82.1 per cent) collaborated with foreign partners for providing technical training while 71.6 per cent of firms had collaboration for importing drawing and design. Roughly half of the firms (50.7 per cent) collaborated for import of machinery. Japan was the most preferred destination for overseas training purposes followed by Germany and Taiwan.

The opinions of managing directors (MDs) towards using ICT were collected on a five-point scale (1 – not important, 2 – to some extent, 3 – important, 4 – very important, 5 – most important). Although MDs gave importance to all the factors for using ICTs as presented in Table 3.8, ICTs provide better management control was cited the most important reason for the adoption of new technologies.

Factors that affect access to ICTs were also measured on a scale of 1 to 5 (1 – not a constraint, 2 – to some extent, 3 – yes, 4 – serious constraint, 5 – severe constraint). Surprisingly the respondents rated them as between 'not a constraint' and 'to some extent'. Speed of communication, lack of physical infrastructure and lack of skilled manpower were considered the three most important factors affecting access to ICTs.

The responses of MDs presented in Table 3.9 suggest that technological infrastructure was not a major problem in the adoption of ICTs in Malaysia. We also solicited data about the consequences of the adoption of ICTs on a five-point scale.

Table 3.10 highlights the consequences of using ICTs. All factors listed in Table 3.10 were considered to be positively influenced except for labour requirements. We can infer from Table 3.10 that adoption of ICTs does not result in reduction in workplaces.

Table 3.8   Reasons for using ICTs

	Mean	(SD)
Provides better management control	3.600	(0.494)
External competition	3.569	(0.558)
Increased in sales per annum	3.569	(0.585)
Efficiency in production	3.569	(0.558)
Reduces per unit cost	3.523	(0.615)
Provides opportunity to produce other related items	3.523	(0.562)
Increased in export per annum	3.492	(0.616)
Flexibility in design	3.477	(0.664)
Enables to produce the product suitable to local conditions	3.446	(0.587)
Internal competition	3.308	(0.528)
Helps to get information about latest markets trends	3.123	(0.451)

Table 3.9 Constraints in accessing ICTs

	Mean	(SD)
Speed of communication	1.862	(0.583)
Lack of physical infrastructure	1.754	(0.791)
Lack of skilled manpower	1.677	(0.752)
Internet subcription fee	1.646	(0.856)
Communication cost	1.631	(0.72)
Non-availability of useful information	1.585	(0.659)
Internet speed	1.415	(0.7480
Availability of Internet connection	1.062	(0.348)

Table 3.10 Consequences of using ICTs

	Mean	(SD)
Lead time advantage	3.400	(0.494)
Mechanisation of small parts assembly	3.369	(0.547)
Flexibility in products design	3.323	(0.471)
Reorganisation and management	3.323	(0.471)
After sales support	3.323	(0.589)
Deskilling	3.308	(0.498)
Productivity gains	3.139	(0.348)
Labour requirements	2.000	(0.000)

Table 3.11 Source of competitiveness

	Mean	(SD)
Market network	3.537	(0.532)
Technology collaboration	3.522	(0.725)
Low wage and salary rates	3.508	(0.587)
Quality of the products	3.493	(0.612)
Delivery schedule	3.493	(0.66)
Brand name/ Good will	3.493	(0.66)
R&D	3.463	(0.611)
Size of your operation/Market share	3.448	(0.61)
Advertisement	3.388	(0.65)
Flexibility of the products	3.373	(0.573)
Low overhead costs	3.343	(0.687)
The proximity of local supplier	3.194	(0.701)

Table 3.11 shows the average opinion of MDs on the sources of competitiveness, measured on a scale of 1 to 5. Assignment of highest importance to market network suggests that traditional sources of competitiveness such as product quality and delivery schedule have lost their pivotal position.

Table 3.12   Mode and effectiveness of learning processes

	Mean	(SD)
Formal training	3.552	(0.681)
Learning by doing	3.313	(0.679)
Learning from technical partner	3.090	(0.733)
On-job training by other institution	2.940	(0.649)
Searching the Internet	2.864	(0.677)
Overseas training	2.470	(0.863)

Results presented in Table 3.11 suggests that non-price sources such as market networking and technological collaboration are more important than low wages and salaries which is a price related source of competitiveness. Finally opinion on effectiveness of learning processes is presented in Table 3.12.

The MDs rated the various modes as poor to average, measured on a scale of 1 to 5, (1 – not effective, 2 – poor, 3 – average, 4 – good, 5 – very good). Surprisingly although most of the firms sent their workers for overseas training, MDs found that formal training is more effective than overseas. Literature is full of findings (Little *et al.*, 1989) that on job training and learning by doing are the most preferred modes of knowledge acquisition in SMEs, the finding of Malaysian SMEs differ from the general perception. One of the possible reasons could be that SMEs in Malaysia may be using more advanced production technologies which require formally trained workers. Formally trained workers may be more effective in adapting and modification of new technologies.

7   Entrepreneurship and the level of ICT adoption

This section presents bivariate analysis of the intensity of ICT adoption with other firm- and industry-specific characteristics. Although it is difficult to measure entrepreneurship, some firm-specific characteristics representing entrepreneurship include MDs education and knowledge base, skill intensity of firms, international orientation of firms, and learning processes. The association between these characteristics and the degree of ICT adoption has been examined in the next sub-section.

7.1   Levels of ICT adoption

ICT intensity is categorized according to three levels, namely: low, moderate, and advanced. This categorization was based on the usage of ICT tools and purposes. Low level ICT adoption refers to basic uses of email, Internet and management information system. These tools are mainly used for communication purposes. Moderate level ICT adoption refers to the uses of ICT tools in the product or process design and also throughout the manufacturing process. Examples of moderate level adoption include CAD/CAM, CAE, FMS, and NCMT. Advanced level ICT adoption refers to uses of more advanced ICT applications that include the uses of web enabled services and portal for supply chain and marketing purposes.

The distribution of firms according to the intensity of ICT adoption suggest that only four firms had low adoption of ICTs (6 per cent), 19.4 per cent were moderate users of ICTs whereas 74.0 per cent of firms were advanced users of ICTs. Interviews conducted later and content analysis of web sites revealed that although 50 respondents had web enabled services, most of these web contents provided basic information regarding firm's product and services, distribution and feedback. These websites also lacked important features such as online purchases, real-time communication and interaction.

7.2   Sectoral pattern of ICT adoption

In this sub-section we examine the determinants of ICT adoption level among sample SMEs by looking at the different sectoral patterns. The sectoral distribution of firms classified by the extent of the use of ICTs is presented in Table 3.13. The table shows that the adoption of ICTs was very much determined by types of industries. Knowledge intensive industries such as chemical and pharmaceutical, electrical and electronics, and hardware and machinery had the tendency to adopt advanced level of ICTs. Only one chemical and pharmaceutical firm was found to have low level of ICT adoption as it was a traditional herbs producer operating on a very small scale. Three food and beverage firms had low level of ICT adoption and they produced ice cube and ice cream.

Table 3.13   Sectoral distribution of sample firms by the level of ICT use

Sector	Level of ICT use			Total
	Low	Moderate	Advanced	
Automotive & transportation		1 (100)		1 [1.49]
Chemical & pharmaceuticals	1 (10.00)	1 (10.00)	8 (80.00)	10 [14.93]
Electrical & electronics		2 (33.33)	4 (66.67)	6 [8.96]
Fashion & textiles		3 (50.00)	3 (50.00)	6 [8.96]
Food & beverage	3 (50.00)		3 (50.00)	6 [8.96]
Furniture			8 (100)	8 [11.94]
Hardware & machinery		6 (25.00)	18 (75.00)	24 [35.82]
Paper & stationery			6 (100)	6 [8.96]
Total	4 (5.97)	13 (19.40)	50 (74.63)	67 [100]

Note: Figures in parentheses are row percentage while is square brackets are column percentage.

Furniture industry in Malaysia is another sector that is moving towards export markets, thus it has become more technology driven. It was found that all firms in this sector adopted advanced ICTs. This trend was not in tandem with the overall SMEs in Malaysia due to the fact that a majority of the respondents were drawn from more advanced/technologically driven industries such as hardware and machinery, chemical and pharmaceutical and electronics industries. They were found to be more open and receptive to survey and interviews.

7.3 Skill intensity and the degree of ICT adoption

Distribution of firms according to the intensity of ICT use and skill intensity is presented in Table 3.14. The table shows that level of ICT adoption is positively associated with the skill intensity of sample firms. This is reflected by the fact that roughly half of the workforce (45.81 per cent) in low level of ICT using firms did not have formal training. On the other hand percentage of such workers in moderate and advanced level of ICT using firms was just 1.60 per cent and 0.62 per cent respectively. Not only that the percentage of technically trained workers in low level of ICT using firms was just 9.69 per cent whereas the percentage of such employees in moderate and advanced ICT using firms was 22.60 per cent and 27.58 per cent respectively. This suggests that the level of ICT adoption is positively related to the skill intensity of firms.

Table 3.15 presents the distribution of firms according to the academic qualifications of Owners/MDs. Although low level of ICT using firms were just four, 50 per cent of them were managed by persons with primary or secondary level of education whereas not a single moderate or advanced ICT using firm was being managed by persons with such qualifications. A comparison of the academic background of moderate and advanced ICT using firms suggests that MDs of the latter were more qualified. For instance, 38.46 per cent of moderate ICT using firms were being managed

Table 3.14   Skill intensity of workforce and the degree of ICT adoption

Workers (in %)	Level of ICT use			Total
	Low	Moderate	Advanced	
Eng. graduates	13.22	19.54	18.16	18.27
PG/Graduate	4.85	8.00	8.19	8.06
Diploma	15.86	27.68	24.48	24.80
Tech. training	9.69	22.60	27.58	26.20
UG	10.57	22.58	20.97	20.61
Others	45.81	1.60	0.62	2.06
Total	100	100	100	100

Table 3.15   MDs' qualification and ICT adoption levels

Workers (in %)	Level of ICT use			Total
	Low	Moderate	Advanced	
Primary	1 (25.0)			1 (1.54)
Secondary	1 (25.0)			1 (1.54)
Diploma	1 (25.0)			1 (1.54)
B. Engineering		3 (23.08)	2 (4.17)	5 (7.69)
B. Admin/Comm		3 (23.08)	10 (20.83)	13 (20.00)
B. Science		1 (7.69)	7 (14.58)	8 (12.31)
MBA	1 (25.0)	5 (38.46)	24 (50.00)	30 (46.15)
M. Engineering		1 (7.69)		1 (1.54)
M. Science			5 (10.42)	5 (7.69)
Total	4	13	48	65

Note: Figures in the parentheses are column percentages while in square brackets are row percentage.

by MBA MDs while percentage of advanced ICT using firms managed by person with such qualification was 50 per cent.

Generally, it has been widely accepted that knowledge intensive or technologically driven industries require highly qualified staff to understand and be able to make informed decisions regarding the various ICT tools and applications in the quest for efficiency and effectiveness. It was found that the educational attainment or qualification of the firm's managing director had a positive relationship with the level of ICT adoption. MDs of 39.58 per cent of advanced ICT using firms had at least a bachelor's degree in engineering, administration/commerce or science.

It was also found that four firms had low adoption of ICTs and all their owners' educational levels were also low, ranging from primary to diploma levels. On the other hand, the educational attainment or qualifications of owners of firms in the moderate and advanced levels of ICT adoption was much more higher, with at least tertiary level education. There is a strong link between technologically driven industries and the qualification of the owners. Knowledge driven firms i.e. chemical and pharmaceutical, electrical and electronics, hardware and machinery, had a very high ratio of advanced ICT adoption levels, which was 80 per cent, 60 per cent, and 75 per cent respectively. In fact, MDs of these firms in these three industries had at least a bachelor's degree in a specified discipline.

7.5   Firm size and ICT adoption levels

Firm size has been measured by the number of employees in this study. Firms were classified according to three major groups namely micro – firms

Table 3.16   Firm size and ICT adoption

Firm use	Level of ICT use			Total
	Low	Moderate	Advanced	
Micro	2 (50.0)	3 (23.08)	5 (10.0)	10 (14.93)
Small	2 (50.0)	9 (69.23)	39 (78.0)	50 (74.63)
Medium		1 (7.69)	6 (12.0)	7 (10.45)
Total	4	13	50	67

Note: Figures in the parentheses are column percentage.

with less than 50 employees, small – firms with between 50 to 150 employees and medium – firms with between 150–200 employees. The distribution of firms by size and the level of ICT adoption is presented in Table 3.16.

Based on the cross tabulation of firm size and the level of ICT adoption, small and medium firms had the tendency to have moderate and advanced ICT adoption. Many firms (90 per cent) that had advanced ICT adoption were in the small and medium-size categories. At the same time, 76.92 per cent of firms that had moderate ICT adoption were in small and medium-size categories. The micro firms had the tendency to have low level of ICT adoption. This could be due to the firms' scale of operation and their inability to bear the ICT cost and also the simple daily operations that could be easily managed without ICTs.

7.6   Owners' age and ICT adoption

It can be seen from the distribution of sample firms by MD's age and degree of ICT adoption presented in Table 3.17 that low level of ICT using firms

Table 3.17   MDs' age and ICT adoption

MDs' age group	Level of ICT use			Total
	Low	Moderate	Advanced	
<35		2 (15.38)	8 (16.33)	10 (15.15)
35–39		4 (30.77)	6 (12.24)	10 (15.15)
40–44		3 (23.08)	11 (22.45)	14 (21.21)
45–49	1 (25.0)	2 (15.38)	12 (24.49)	15 (22.73)
50–54	1 (25.0)	1 (7.69)	8 (16.33)	10 (15.15)
55–59	1 (25.0)		4 (8.16)	5 (7.58)
60 +	1 (25.0)	1 (7.69)		2 (3.03)
Total	4	13	49	66

Note: Figures in the parentheses are column percentage.

were being managed by older persons compared to moderate and advanced ICT using firms. It is quite possible that MDs of low level of ICT using firms were less technology savvy and hence they were not sure about the potential benefits of ICTs and did not favour for modern tools. On the other hand, younger MDs, being more familiar with potential benefits, adopted more advanced tools.

It can be seen from Table 3.17 that all MDs of the low level ICT using firms were over 45 years old whereas 84.5 per cent MDs of moderate and 75.5 per cent MDs of advanced level of ICT using firms were less than 50 years old.

7.7   Mode of learning and ICT adoption

Table 3.18 shows that 56 SMEs provided overseas training to their employees, 47 (83.92 per cent) of these firms were in advanced ICT adoption category. We can generalize that firms in advanced ICT adoption category have the tendency to provide external training to their employees. Interviews conducted revealed that these firms had to conform to stringent requirements of their foreign customers or suppliers. Reasons given include the need to understand new customer requirements, to learn new technologies and processes and to source for new machinery or training provided by machinery suppliers. Popular destinations are Japan, Germany, Taiwan and Korea.

Thirty-seven firms provided formal training and 30 provided on-the-job training. The ratio of formal training to on-the-job training was higher in advanced ICT adoption categories. This could be due the fact that advanced ICT adoption needs more formal training especially in utilizing ICT for logistics, marketing and customer services. In contrast, low ICT adoption category firms were more labour-oriented, thus on-the-job training was the preferred choice.

Table 3.18   Learning processes and ICT adoption

Type of training	Level of ICT use			Total
	Low	Moderate	Advanced	
External				
Local	1 (50.0)	5 (38.46)	3 (6.0)	9 (13.85)
Overseas	1 (50.0)	8 (61.54)	47 (94.0)	56 (86.15)
Internal				
Formal	1 (25.0)	4 (30.77)	32 (64.0)	37 (55.22)
On Job	3 (75.0)	9 (69.23)	18 (36.0)	30 (44.78)
Total	4	13	50	67

Note: Figures in the parentheses are column percentage.

8   Statistical analysis

This section has two major parts. The first part highlights the analysis of variance of factors that are expected to influence the extent of the use of new technologies among sample firms. In second part we present multivariate analysis of the adoption of ICTs. The variables that have been included in the analysis and their measurements are discussed in the following sub-section.

8.1   Specification of variables

Most of the variables used in the analysis are opinion variables. Opinions of MDs were sought on a five-point scale. The information on the type ICT tools was collected and subsequently firms were grouped into three categories based on the intensity of ICT adopted. In addition, few binary variables have been used in the analysis. We have not been able to examine the impact of performance variables on the adoption of ICTs due to lack of data. The variables used in the analysis were measured and defined as follows:

8.1.1   Level of ICT adoption (IT_LEVEL)

X_i refers to the level of ICT adoption, Where X=1 advanced level of ICT adoption was present and X=0 when low and moderate level of ICT adoption was present. However in univariate analysis (ANOVA), variables have been analysed keeping low level of ICT using firms as a separate category. In logit analysis, it became necessary to merge low and moderate category of firms into one category because very few degrees of freedom in the lower category of firms did not permit to estimate parameters of logit model. The level of ICT adoption is expected to have associations with the following factors.

8.1.2   Managing Director's qualifications (MDEDU)

Although data on actual qualification of MDs were collected, they were quantified into three categories for the purpose of statistical analysis. MDEDU was assigned vales one to those MDs who had formal qualification of below tertiary level. Value 2 was assigned to bachelor's degree holder MDs while master's degree holders were assigned highest value, i.e. 3.

8.1.3   Skill intensity (SKILL)

Skill intensity of firm was measured by the ratio of sum number of engineering graduates and postgraduate/graduates to the total number of full-time employees. Firms with high skill intensity ratio is expected to have advanced level of ICT adoption.

8.1.4   Benefit of ICT use (FLEX)

Several benefits such as augmentation in competitiveness, reduction in production costs, increase in sales turnover and exports, ability of ICTs to contribute to flexibility in product designs have been cited in the literature. We have used flexibility in product designs as benefit of ICT use. It was measured on a five-point scale ranging from 'does not augment flexibility' to 'most important technology for manufacturing products of flexible designs'. It is expected that production technologies such as CAD/CAM using firms assigned more importance to this aspect of ICTs.

8.1.5   Firm size (SIZE)

Sales turnover and number of workers have been used as proxies of size in the literature. Since most of the MDs had reservations in providing financial data of their firms, we had to use number of workers as proxy of size. It is expected that larger firms are users of more advanced ICTs because they are in better position to exploit the potentials of new technologies. A comparatively larger firm is also more likely to have a higher level of ICT adoption rate because ICT cost is spread thinly considering their large production scale.

8.1.6   Foreign collaboration (FORN_COL)

This was measured on a binary scale. It was assigned vale 1 if firm had a foreign partner and 0 otherwise. In case of Malaysia, it is quite possible that firms with foreign collaboration might have adopted more advanced ICTs. This is because fairly large number of sample firms (44.78 per cent) belonged to electronics, and hardware & machinery sector. These firms were basically component manufacturers of MNCs. Hence technological collaborations might emerge as a determinant of ICT adoption.

8.1.7   Mode of knowledge acquisition (FORM_TRN)

There are several ways of knowledge acquisition such as learning by doing, on job training, learning from technical partners, overseas training, and formal training provided by technology institutions which is a prerequisite of effective use of new technologies. We sought the opinion of MDs on the preferred mode of learning on a five-point scale. It is expected that opinion on formal training might have been given more importance by the advanced users of ICTs because such tools need customization and that can be successfully done by formally trained staff.

8.1.8   Adequacy of learning opportunities within industrial cluster (ADE_TRN)

This variable was measured on a binary scale. Firms were assigned value 1 where MDs opined that adequate learning opportunities within the cluster are necessary and 0 otherwise. Continuous upgrading of knowledge is essential for effective use of ICTs because product cycle of ICT tools is very small and they become obsolete very frequently. In order to use new tools efficiently, continuous upgrading of skills of worker is imperative and this can be easily achieved if the training facilities are available within close vicinity of firms. Therefore, advanced users of ICTs might have given more importance to availability of learning opportunities within industrial clusters.

8.1.9   Sources of competitiveness (MARK_SHR)

Competitiveness in the market included the quality of the product, flexibility in design, advertisement, R&D, size of operation/market share, delivery schedule, low overhead costs, brand name/goodwill, market network, and low wage and salary rates. We expect that the role of market share might be a more important source of competitiveness. The variable was measured on a five-point scale ranging from 1 as no role to 5 as most important role.

8.1.10   Consequences of ICT use (PROD)

The consequences of using ICTs included productivity gains, deskilling, labour requirements, lead-time advantage, flexibility in products designs, reorganization and management, mechanization of small parts assembly, and after sales support. Productivity gains are expected to be considered most visible and much needed consequence of the ICT adoption. This is also an opinion variable. Opinions of MDs about the productivity gains as a result of adoption of ICTs were sought on a five-point scale. This opinion is expected to have positive relationship with the likelihood of advanced level ICT adoption.

8.2   Analysis of variance

To investigate the associations between selected respondents' characteristics and the adoption of ICTs, analysis of variance (ANOVA) of variables discussed above is presented in Table 3.19.

Table 3.19 shows that intensity of ICT adoption does not differ significantly among different size of firms. Similarly opinion on productivity gains by MDs of different levels of ICT using firms does not differ significantly. Surprisingly MDs of varying levels of ICT using firms did not think that augmentation in flexibility was a major benefit of ICT use. The results regarding other variables are according to our expectations.

8.3   Logit analysis

Theoretical framework used in this study is presented in Figure 3.4. It can be seen from the figure that the intensity of the adoption of new technologies is influenced by external as well as internal factors. Arrows in the diagram show the direction of causality.

A basic model was specified to estimate the determinants of high level ICT adoption. The binary logistic regression function was chosen in this analysis because the dependent variable was a dichotomy and the independent variables were complex, ranging from nominal, ordinal and interval scales. In this study, logistic regression was used to predict the level

Table 3.19 Analysis of variance

Variables	IT-level	Mean	F. value	Sig.
1 Ratio of employees with master and bachelor degree to the total employee (SKILL)	Low	0.10	6.055***	0.004
	Moderate	0.28		
	Advanced	0.29		
2 Education of MDs (MDEDU)	Low	1.50	7.664***	0.001
	Moderate	2.50		
	Advanced	2.60		
3 Foreign collaboration (FORN_COL)	Low	0.50	4.789**	0.012
	Moderate	0.77		
	Advanced	0.94		
4 Benefits of ICT use (FLEX)	Low	3.50	0.865	0.426
	Moderate	3.70		
	Advanced	3.42		
5 Firm size (SIZE)	Low	56.75	1.340	0.269
	Moderate	110.62		
	Advanced	121.07		
6 Effectiveness of learning processes (FORM_TRN)	Low	2.25	9.896***	0.000
	Moderate	3.62		
	Advanced	3.55		
7 Sources of competitiveness (MARK_SHR)	Low	2.50	5.922***	0.004
	Moderate	3.54		
	Advanced	3.50		
8 Consequences of ICT use (PROD)	Low	3.00	0.167	0.847
	Moderate	3.15		
	Advanced	3.14		
9 Adequacy of learning opportunities (ADE_TRN)	Low	0.25	4.833**	0.011
	Moderate	0.69		
	Advanced	0.98		

Note: **, *** significant at 5 %, and 1 % respectively.

of ICT adoption and to determine the percentage of variance in the level of ICT adoption explained by the independent variables; to rank the relative importance of independents; and to understand the impact of covariate control variables. The logistics regression does not assume linearity of relationship between the independent variables and the dependent variables, does not require normally distributed variables, does not assume homoschedasticity and in general has less stringent requirements.

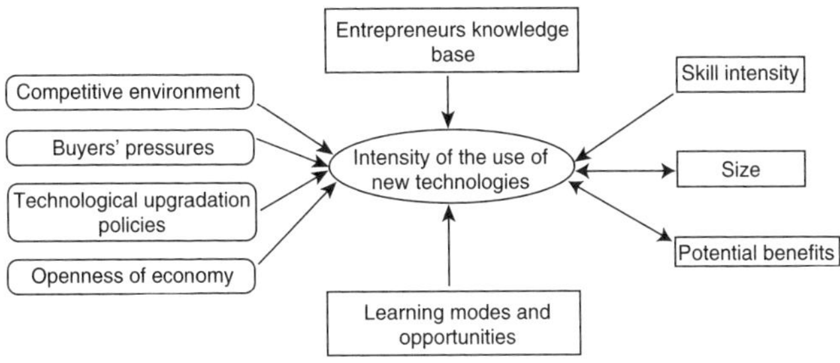

Figure 3.4   Theoretical framework

The binary logistics function provides estimates that must lie in the range between 0 and 1. For logistic model, the explicit form would be:

$$p(x) = \frac{1}{1+e^{-(\beta_0+\beta_1 MDEDU+\beta_2 SKILL+\beta_3 FORN\_COL+\beta_4 FLEX+\beta_5 FORM\_TRN+\beta_6 ADE\_TRN+\beta_7 PROD+\beta_8 SIZE+\beta_9 MARK\_SHR)}}$$

In logit form, it would be

$$\ln[P/(1-P)]$$
$$= \beta_0 + \beta_1 MDEDU + \beta_2 SKILL + \beta_3 FORN\_COL + \beta_4 FLEX + \beta_5 FORM\_TRN$$
$$+ \beta_6 ADE\_TRN + \beta_7 PROD + \beta_8 SIZE + \beta_9 MARK\_SHR$$

We chose to apply maximum likelihood estimation for estimating the parameters in the model since it requires no restriction on the characteristics of the independent variables. Hypotheses are:

$$H_0 : \beta_i = 0$$
$$H_1 : \beta_i \neq 0$$

Logit analysis results are presented in Table 3.20. As can be seen from the table that four different models have been tried. This was necessary to control for multicollinearity. In Model I all the variables except those that were insignificant in ANOVA were specified while in other models MDEDU, SKILL, and FORM_TRN were specified individually.

Results presented in Table 3.20 are similar to what we found in ANOVA except in the case of the role of market share variable. This was highly significant in univariate analysis but did not emerge as important factor in determining the intensity of the adoption of ICTs in multivariate analysis. This is primarily because of treatment of dependent variable. In ANOVA three categories of ICT adoption were considered whereas in Logit it was not possible to do so.

Table 3.20   Logit analysis

Independent Variables	Dependent variable: IT_LEVEL							
	Model I		Model II		Model III		Model IV	
	Coeff.	Z	Coeff.	Z	Coeff.	Z	Coeff.	Z
Intercept	−4.176		−1.233		−0.117		−1.328	
MDEDU	0.342	0.454	0.968	1.953*				
SKILL	−0.031	−0.009			4.515	1.702*		
FORN_COL	1.879	1.891*						
FORM_TRN	0.452	0.747					0.690	1.716*
MARK_SHR	−0.123	−0.162						
ADE_TRN	2.176	2.637**						
Observations	64		65		67		67	
Log likelihood	−27.734	[0.011]	−34.28373	[0.046]	−36.412	[0.080]	−36.419	[0.0802]
Classification power of the model	81.35		80.00		79.10		77.61	

Note: Figures in square brackets are level of significance of the function; *→ 10% and **→ 1% level of significance.

Statistical analysis suggests that formal education and knowledgebase of MDs and skill intensity are very crucial for the adoption of advanced technologies in SMEs. These findings do not merely meet our expectation but are in tandem with the existing literature (Lal, 2002; Pohjola, 2001; Doms *et al.*, 1997). Doms *et al.* (1997) argued that new technologies led by ICTs are by and large skill biased whereas Lal (2002) in his study of SMEs in India found that formal education of MDs and skill intensity significantly influenced the adoption of e-business technologies.

The unique contribution of the study lies in its ability in capturing the role of learning modes and opportunities in the adoption of new technologies. Results presented in Table 3.20 suggest that formal training has been assigned more importance than preferred modes in SMEs such as learning by doing, and on-job training. Another uniqueness of the study is in identifying the critical role of adequacy of knowledge acquisition opportunities within industrial cluster. The study finds evidence of technological collaboration being accorded an important role in the adoption of new technologies.

In many other developing countries, where technological profile of SMEs is not very strong, the entry level of ICTs is usually used. As a result skill intensity may be irrelevant because workers do not require high levels of skills to operate tools like MIS. The finding of Oyelaran-Oyeyinka and Lal (2005) is a case in point. The authors compared type of ICTs used in SMEs in India, Nigeria, and Uganda. It was found that technological profiles of SMEs in three countries differ significantly. Although the findings are to some extent similar to existing literature, country-specific characteristics can clearly be seen as having their impact on the adoption of SMEs in Malaysia. Emergence of skill intensity as a significant factor in influencing the intensity of ICT use in Malaysia suggests that ICT used in Malaysian SMEs are more complex than other developing countries. A strong and positive correlation between foreign technological collaboration and the degree of ICT adoption suggest that SMEs in Malaysia target global competitiveness and produce international quality of products.

The study finds two surprising results. First, the insignificant role of size of firms in the adoption of ICTs and second, MDs did not consider product quality as a major source of competitiveness. This is contrary to the expectation and differs from the findings of earlier studies (Doms *et al.*, 1997; Lal, 1996).

9   Conclusion and policy implications

Although findings showed that the majority of the respondents had advanced ICT adoption rate. However, results cannot be generalized because majority of sample size were drawn from more advanced or technology driven industries such as hardware and machinery, chemical and pharmaceutical and electronics. It is generally known logic that techno-logically advanced industries should have higher levels of ICT adoption rate. This sentiment was echoed by many representatives from both government and SMEs associations in Malaysia and including SMEs owners themselves.

One more important finding is that educational attainment of companies' directors had a bearing on ICT adoption. There is also a strong link between technology driven firms and the qualification of their leaders. Collaboration with foreign partners was found to be a usual practice among SMEs in Malaysia. Follow up interviews also attested to this phenomenon. A president of an SME association in Malaysia concurred that SMEs in this country normally employed workers over 40 years of age and most of them could only speak and write Chinese. English proficiency was recognized as the main problem, thus hindering training or collaborative efforts with foreign partners.

It is clear that the Malaysian government is serious about enhancing SMEs competitiveness in response to growing threats of globalization. However the policies introduced two decades ago were very much in line with efforts to increase participation of Bumiputra (natives), sometimes at the expense of overall development. It is also interesting to note that the Malaysian government has begun to recognize the importance and contribution of SMEs to Malaysian economy with the establishment of SMEs Development Bank in 2006.

The findings of the study suggest that apart from providing financial support to SMEs Malaysian government need to focus on human resource development policies. It is twofold. First, government needs to establish technology institutions that can provide job-oriented formal training. And second, short-time training opportunities near the workplaces may be very helpful for skill up-gradation of workers. Provision of such opportunities is expected to contribute to efficiency and high productivity of workers. We have not been able to evaluate the impact of ICT on economic performance of sample SMEs due to the lack of data on performance indicators.

4

Discriminants of the Adoption of ICTs in Central America: A Case Study of Costa Rican SMEs

1 Introduction

Small and medium-sized enterprises (SMEs) are considered as the backbone of economic activity in any country. In developing countries efforts have been made to increase their competitiveness and to achieve a successful international trade insertion. However, structural problems in these countries severely affect the performance of SMEs – the absence of adequate Information and Communication Technologies (ICTs) is one of the most recurrent bottlenecks found in developing countries.

The ICTs are key instruments for achieving a solid, sustained competitiveness in both the internal and external markets. New technologies can help SMEs to increase earnings by allowing them to enter into new markets and achieving competitive advantages. Given the highly positive externalities derived from these technologies, it is frequently suggested that their development is a process that should go hand-in-hand with the government and non-public agencies.

For Costa Rica, a small open economy with less than 5 million inhabitants, the former considerations are also true. With more than 90 per cent of its firms as SMEs (excluding micro-firms), the country achieved substantial progress in terms of basic telecommunications that heavily supported the established firms. However, during the last few years, the country has lagged in the evolution of modern ICTs, affecting the performance of SMEs and limiting their access to bigger markets and new businesses. The analysis presented here is a clear example that ICTs are important tools for the identification and consolidation of opportunities but at the same time, the absence of adequate infrastructure can limit the expansion of SMEs in the country.

The general objective of this chapter is to examine the level of use of ICTs among SMEs in Costa Rica. The study surveyed a sample of 68 manufacturing SMEs in Costa Rica and visited several firms, government institutions and industrial organizations to evaluate the impact of ICTs in the productive processes as well as to discuss future plans to expand ICT infrastructure.

The organization of the chapter is as follows. Section 2 provides a historical overview of the importance of SMEs in the Costa Rican economy between 1960 and 2004. It also reviews the most important industrial programmes applied during the same period. Then, Section 3 evaluates the results of the sample, emphasizing key relationships between ICTs and managerial variables and the perception of businessmen on the importance of ICTs to their companies. Section 4 discusses statistical results based on the data collected from survey of firms. Case studies of three advanced ICT using firms are presented in Section 5. Finally, summary and conclusions are presented in Section 6.

2   SMEs background

This section depicts the experience of Costa Rican SMEs over the last 20 years and analyses changes in SMEs-oriented public policies and their economic performance. First of all we deal with SMEs definition and the criteria used to define what an SME is. Then we describe the evolution of the Costa Rican SMEs from historical point of view using a set of economic indicators and sectoral characteristics. Subsequently SMEs' contributions to the economy during the last 20 years are analysed. Finally, a review of different SMEs-oriented public programmes defined by different governments during the last two decades is presented.

2.1   The definition of SMEs

Along with micro-enterprises, SMEs represent the vast majority of firms in Costa Rica. In total, these firms generated 263,611 jobs in 2003 and represented 96 per cent of total private sector firms. According to Gómez (2002), 78.8 per cent of 14,015 SMEs are small firms and 21.1 per cent are medium enterprises. Although SMEs represent an important part of the Costa Rican economy, it was recently that an official SMEs definition was acknowledged.

The Ministry of Economy, Industry and Commerce (MEIC) has given the official definition of SMEs in Costa Rica, although other SMEs-related institutions have their own definition. The official concept of SMEs used by the MEIC considers elements such as number of employees, annual sales and technological elements surrounding the enterprise. Gómez also uses this definition in his study.

The official criteria used to categorize SMEs in Costa Rica are shown in Table 4.1. In general, an SME is a firm that has between six and 100 employees, with annual sales between $ 500,000 and $ 1,000,000 and investments in machinery and equipment ranging between $ 250,000 and $ 500,000.

Table 4.1   Criteria used to define SMEs in Costa Rica

Size	Characteristics	Values (in US dollars)
Small enterprise	Number of employees	6–30
	Investment in machinery, equipment and tools (maximum)	$250 000
	Annual sales	$500 000
Medium enterprise	Number of employees	31–100
	Investment in machinery, equipment and tools (maximum)	$500 000
	Annual sales (maximum)	$1 000 000

Source: Castillo *et al.* (2003).

Statistics for the total number of firms in Costa Rica are provided by MEIC and the Costa Rican Social Security Institute. The total number of companies was 74,833 by the year 2002 of which 3,156 were medium-sized (4 per cent) and 11,711 were small firms (16 per cent), while SMEs accounted for 14,866 firms (20 per cent). The remaining 80 per cent were mostly micro and large firms. Micro enterprises in Costa Rica are defined as firms where total workforce is limited to ten persons with an investment limit of $ 250,000 in machinery and equipment, and sales turnover not exceeding $ 500,000.

2.2   Historical background

This section provides a historical perspective on the importance of the manufacturing sector in the Costa Rican economy, emphasizing the contribution of SMEs in total output, employment, exports and imports. Despite the importance of SMEs, historical information is, however, hardly available.

The evolution of the Costa Rican manufacturing sector can be analysed over three sub-periods. The first stage, before 1960, is characterized by the predominance of the agricultural sector in the overall economy and the poor development of the manufacturing structure. During this period, two crops, coffee and banana, represented most of the exports. Then, during the second period, between 1960 and 1985, the Industrialization Model (IM) was actively implemented and consequently the manufacturing sector emerged as the engine of growth of the Costa Rican economy (6 per cent average growth rate experienced between 1960 and 1970). These first two periods comprise the pre-Structural Adjustment era. Finally, in the third stage, 1986 onwards, government adopted structural adjustment programme, trade liberalization, and competitiveness policies. In this period, although policy orientation mainly focused on the transformation of SMEs into externally competitive industries, the share of the manufacturing sector in GDP practically remained constant.

Under the agriculture-based phase, few products, particularly coffee and bananas, comprised the core of the national GDP and their share in total exports was over 60 per cent. The dependence on two crops made the country highly vulnerable to fluctuations in prices and thus to external crisis in the balance of payments. Despite this fragility, practically no industrial policy was implemented during this period and the manufacturing sector was a series of small-dispersed firms with no intra and inter-sectoral links and poorly developed managerial capabilities.

Based on the ideas developed by the structuralist theory, spread in Latin America by the Economic Commission for Latin America and the Caribbean, the country adopted a new law, the well-known Act of Industrial Protection and Development (1959) that created a series of tax exemptions, subsidies, and tariffs to spur the development of the local industry. Later, with the incorporation of Costa Rica in the Central American Common Market, the adoption of the Central American Common Tariff provided the package of incentives by providing extra protection to the local infant industry and a bigger market for selling its products. The new legal and institutional framework provided highly positive results. As presented in Table 4.2, the manufacturing output grew at 8.6 per cent annually during the first two decades and increased its share from 14.6 per cent of GDP in 1960 to 20.3 per cent in 1980. In this period, the manufacturing sector was the most dynamic sector of the Costa Rican economy.

Table 4.2   Manufacturing production: growth rates, 1960–1982

Period	Average growth rate (%)
1960–1969	8.9
1970–1979	8.3
1980–1982	–3.7

Source: Pacheco (1998).

Table 4.3   Overall performance of the industrial sector, 1960–1979

	1960–1964	1965–1969	1970–1974	1975–1979	Average
Share in GDP	14.6	17.2	18.8	19.3	17.5
Share in employment	14.4	13.2	12	15.5	13.8
Share in total exports	7.4	20.8	25.4	27	20.1
Share in total imports*	25.5	37.9	40.4	45.7	37.4

Note: *→ Corresponds to 1960, 1965, 1970 and 1975.
Source: Muñoz (2002) and Pacheco (1998).

Data on the contribution of the manufacturing sector to national employment and wages also reflect a significant dynamism. No homogeneous information exists on the contribution of manufacturing employment to total employment. For instance, in Muñoz (2002) the contribution to total employment ranged from 14.4 per cent to 15.5 per cent (Table 4.3), while in Pacheco (1998) it increased from 17 per cent to 20 per cent. It is clear from both the studies that upward trend in employment was experienced by the manufacturing sector, although several concerns have been raised on the asymmetry between the dynamics of the manufacturing output and the pace of job creation in the manufacturing sector. Finally, in terms of remunerations, the sector contributed on an average 15 per cent of the total.

Table 4.3 shows that the share of the manufacturing sector exports and imports experienced a positive growth rate. The share of manufacturing exports almost quadrupled while imports almost doubled their contribution. The Central American market became the most important destination by absorbing 70 per cent of the manufacturing exports. This market was so important that, if we take it off from the analysis, the results show that only 3 per cent of the small-scale production, 6 per cent of the medium-scale output and 8 per cent of the big firms' production were exported outside Central America.

Several structural problems and an obscure international context eroded the IM. So by the end of the 1970s the economy experienced some problems. The 1981–82 crisis forced adoption of a set of policy measures to stabilize the economy and to progressively implement a new development model. After 2 years of stabilization efforts, the country actively began the implementation of trade liberalization and pro-competitiveness policies as part of a broader package of structural adjustment measures.

The Structural Adjustment Programme comprised a set of public policies aiming at reducing and eliminating those barriers that do not allow full utilization of the economic resources. Part of the efforts was addressed to implement a new trade model characterized by lower tariffs and export promotion. By lowering tariffs, the industrial sector would face an increasing competition from international firms, improving its level of competitiveness. By promoting exports, the sector would have enough incentives for diversifying its products. The application of fiscal subsidies and the creation of a wide range of trade-related institutions were the two most important tools used by different governments to execute the shift from one commercial regime to another. The results of this change in the evolution of the manufacturing sector are discussed in the next sub-section.

2.3   The contribution of SMEs to the Costa Rican economy

Data on the importance of SMEs (general and manufacturing) in the Costa Rican economy are scarce. Most of the documents containing information

Figure 4.1 Composition of small and medium enterprises by sector (2000)

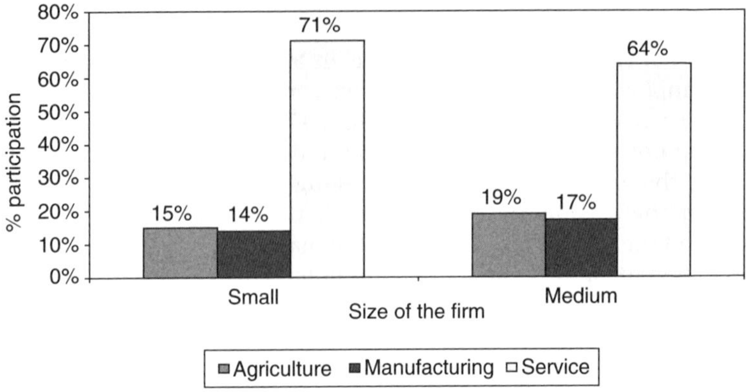

Source: Castillo and Chaves (2003).

about SMEs are from the 1990s and only a few of them make any reference to the present decade. In fact, official information is available until 1998; after this year, only a few sporadic estimations done by public institutions and researches are available.

The total number of SMEs in Costa Rica was 14,816 firms, or 19 per cent of the total number of business that existed by 2000.[5] If no micro firms are considered, SMEs represent 98 per cent of the manufacturing sector. By economic activity, 71 per cent of the small firms belong to services and only 14 per cent to the manufacturing (Fig. 4.1). A similar situation is observed among medium-sized firms, where 64 per cent of them are services-oriented companies while only 17 per cent were engaged in manufacturing activities.

Based on information from Aguilar *et al.* (1998), we estimate the contribution of SMEs to manufacturing output and industrial internal sales during 1990–1997 (Fig. 4.2). The results show that SMEs represent, on average, 33.5 per cent of the manufacturing output and 30.3 per cent of the industrial internal sales during the referenced period. At the national level, the manufacturing SMEs contribution is 7 per cent of GDP.[6]

The share of the SMEs in national and manufacturing employment is presented in Figure 4.3. At the sectoral level, SMEs account for 28 per cent of the employment in manufacturing sector in 1997. Given that the manufacturing sector comprises 15.6 per cent of the national employment, the manufacturing SMEs contribute 4.37 per cent to the total national employment.

---

5 It includes micro-firms, those ones with 5 or less than 5 workers.
6 The average contribution of the manufacturing sector to the Costa Rican GDP is 21 per cent.

Figure 4.2   Contribution of SMEs to manufacturing sales and gross product output

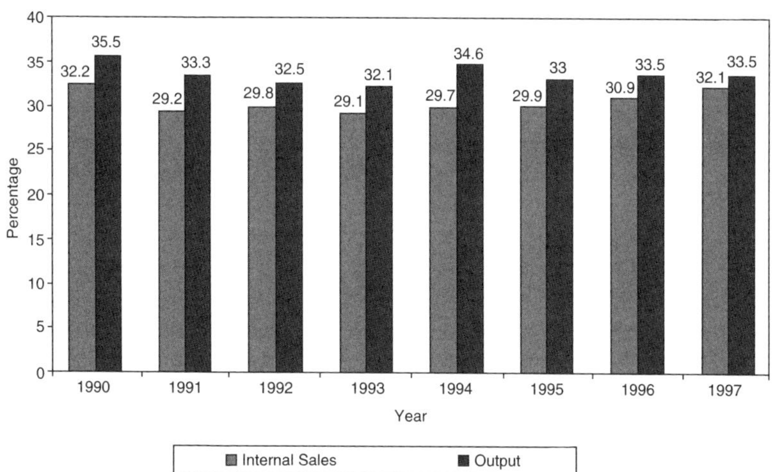

Source: Aguilar *et al.* (1998).

One of the key features of the SME employment dynamics is the declining trend over time. In that sense, while in 1986 the SMEs employed 41,100 workers (38 per cent of the industrial employment), 10 years later their share decreased to 28 per cent (38,800 workers). This situation highly contrasts with the share of big firms in employment: from 61,000 (56 per cent) to 96,000 (67 per cent) in the period of reference.

Figure 4.3   Composition of employment in manufacturing sector

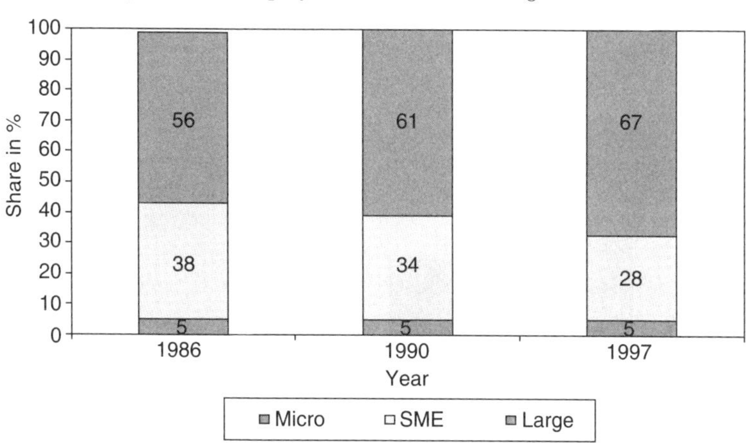

Source: Aguilar *et al.* (1998).

Figure 4.4 Average size of small and medium enterprises

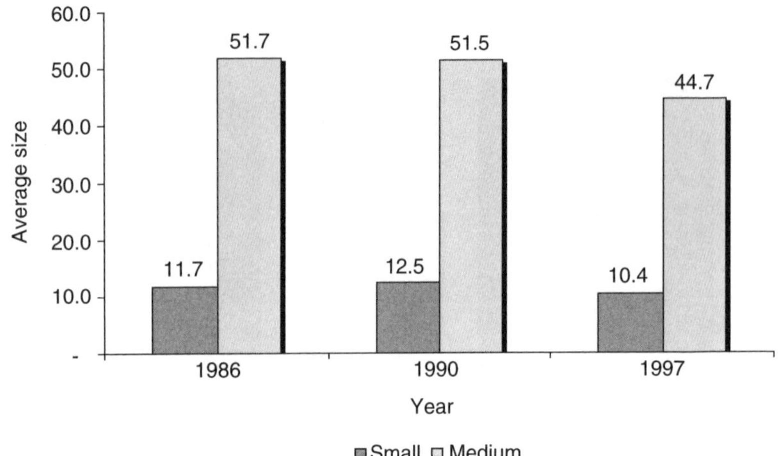

Source: MEIC (1998).

Figure 4.4 shows that the average size of a typical SME has changed over time. For instance, in 1986 a typical small firm had 11.7 employees, while in 1997 it had 10.4. Recent estimations suggest that the decline has continued over time. On an average, a small firm has now nine workers. Similarly,

Figure 4.5 Employment in two most dynamic export sectors SMEs (2004)

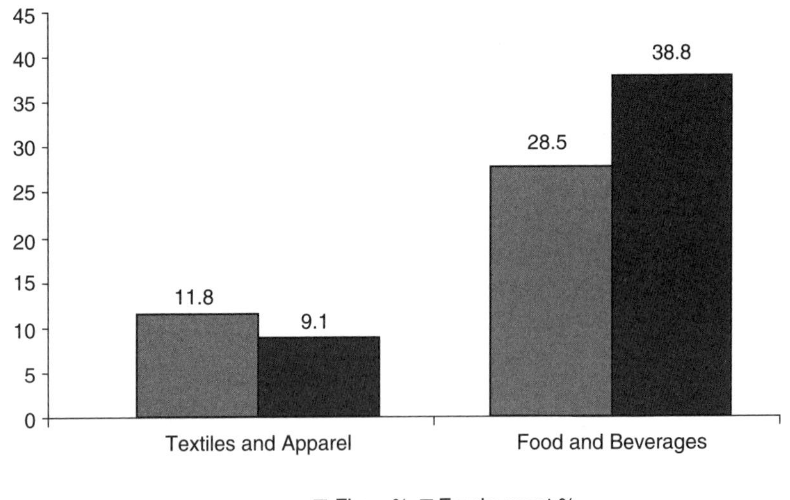

Source: MEIC (2005).

a typical medium-scale enterprise used to have 51.7 workers in 1986 and 44.7 employees in 1997, experiencing the same trend as the other group.

Within the manufacturing sector, most of the labour is concentrated in one group, i.e., food and beverages that comprises almost 40 per cent of manufacturing employment (Fig. 4.5). Textiles and apparel, on the contrary despite its labour-intensive nature, account for only 9 per cent of manufacturing employment. The former diagnosis goes hand-in-hand with recent problems in the apparel and textile sector, product of the North American market contraction, due to the fact that most exports from these sectors go to the USA.

On an average, the SMEs comprised 29.8 per cent of the manufacturing exports during 1990s (Fig. 4.6). Their share increased from an average 27.8 per cent during (1990–1993) to 32.8 per cent during (1994–96). This increase in the share of exports of manufacturing SMEs during 1990s is an evidence of the significant dynamism shown by this group. Given that 76 per cent of the national exports are from manufacturing, on an average the manufacturing SMEs contribute 22.67 per cent to the total Costa Rican exports.

2.4   SMEs competitiveness policy during the structural adjustment period

The incentives, subsidies and industrial programmes applied in Costa Rica during the structural adjustment period, have been conceived along with the need for these firms to make a successful incursion in the international markets. But some of the institutions inherited from the previous regime prevailed and they were adapted to address the problems faced today by these firms. Under the new paradigm, public efforts were made to enhance

Figure 4.6   SMEs contribution to manufacturing exports

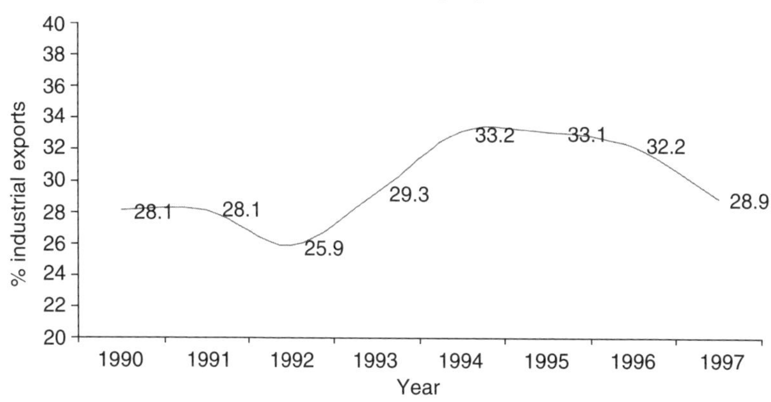

Source: Aguilar *et al.* (1998).

the sector's internal and external competitiveness. With this in mind, a set of policy measures was undertaken.

The first of these actions was the foundation of the Centre for Export Promotion (CEMPRO) and the establishment of the Tax Certificate Allowance (CAT)[7] a subsidy that works as an incentive to exporting firms. Another institution, National Council for Science and Technology (CONICIT) was created with the purpose of promoting research and development and the use of new technologies in the firms. The following section summarizes the most important SMEs-oriented public programmes adopted by different governments in the last 20 years. Programmes are presented in chronological order.

2.4.1 Industrial Transformation Programme (ITP) (1986–1990)

The ITP aimed at creating a stable macroeconomic environment, with equilibrated commercial and budgets deficits, and inflation. The general objective was to create trust among investors, so they can devote themselves to transform their firms into competitive, export-oriented companies in a context of a predictable, less-risky economy. Along the stable macroeconomic situation, this programme also included support policies to increase the benefits of stability. Three sets of actions were designed: one at the macroeconomic and infrastructure realms; a second one is related with policies at the industrial level (meso policies) and the third one directly oriented to firms (micro level).

The programme includes marketing, information acquisition, human resource development, technological management, financing, and international cooperation. Under this programme National System of Scientific and Technological Information was created for the latest information on technological developments in the North and on market trends in the domestic as well as international markets. Training programmes aimed at improving the management, marketing and productive skills were conducted for entrepreneurs as well as workers. The programme stimulated the searching of international agreements in order to partially finance the industrial transformation process and to let firms know the best practices in different fields.

2.4.2 Industrial and Agro-industrial Modernization Programme (IAMP) (1990–1994)

The IAMP was approved in December 1991, established by executive decree in July 1992 and executed in 1993. The programme's general objective was promotion of industrial and agro-industrial sectors, allowing them to reach efficiency levels in local and foreign markets, with the state acting as a

---

7 This subsidy will be explained further in the next section.

facilitator and not interfering in economic activities. In accordance with its orientation, the IAMP mostly defined 'horizontal' measures, that is, no level-specific (meso or micro) measures were considered but broad actions, which aimed at improving the environment to make businesses successful. Main focus of this programme was to foster link between firms and technological innovations in universities and other research institutions. Special attention was given on the promotion of quality standards among firms in order to achieve international standards in their productive processes.

2.4.3   *The Industrial Modernization Programme (IMP) (1994–1998)*

The general objective of the programme was the enhancement of the SMEs competitive capacity (national and international) through technological innovation, management and organizational changes and skillful human resources. As in the previous programmes, the proposed measures did not permit any kind of protection and subsidies, giving the state a new role, more active and involved in decision-making as well as promoting information accessibility. The key tasks assigned to the government were a) reorganization of public services; b) elimination of internal distortions that directly affect costs and competitiveness; c) creation of a smart link between local and foreign markets and d) promotion of modern managerial practices among private sector firms.

2.4.4   *SMEs decree and investment attraction programme (1998–2002)*

Perhaps the most important effort conducted during this administration to support SMEs was the 'National Strategy for the Costa Rican Small and Medium Firm: 1998–2000'. This plan was aimed at firms with earnings below $ 500,000 a year and its policy target was the facilitation of credit access to promote SMEs formalization. Additionally it introduced gender perspectives in SMEs functioning[8] due to the existing link between these firms and household economy as income generating activities. In other words, it was not only an industrial agenda but also a poverty alleviation programme, being the first one in the country that included poverty issue. The agenda proposed by the government was part of a broader programme to attract foreign investment, and the focus of the efforts possibly lied on trade and not on modernization as such. As part of these efforts, during 1999, the National Bank of Costa Rica (the biggest public financial institution in the country) started the 'Financial Services Programme for the Micro and Small Firm' with the sole purpose of satisfying SMEs credit needs while providing financial training services to strengthen the customer-institution relationship trough a financial training programme.

---

8   Most SMEs in Costa Rica are run by male managers as it will be shown in the Data
   Analysis section.

Table 4.4   Institutions and areas of assistance for SMEs

Area of interest	Government	Private initiative
Financial aid	BPDC-BNCR-BCR	FUCODES-ACORDE-MICIT
Training, consulting and technical assistance in management	INA, PROCOMER	FUNDES-CICR-INCAE-UCCAEP
Training, consulting and assistance in quality and environment	INA-UCR-ITCR-MICIT-UNA	CEFOF-CEGESTI-CICR-CACIA-INTECO
Research and development	UNA-UCR-ITCR-MICIT	CAATEC-CRECEX
Web information	MEIC-PROCOMER-UCR	CACIA-INCAE-CADEXCO-FUNCENAT-FUNDES
Marketing	PROCOMER	CAPROSOFT-CICR

Source: http://www.meic.go.cr/; Note:→ Expansion of acronyms is presented in Appendix Table 4.4.

Although a new law 8262 for improving performance of SMEs has been activated in 2002, its impact may not be visible during the period of this study. Hence the provisions are not discussed. Along with government projects, there are also a lot of ongoing programmes related to the development of SMEs and enforced by the private sector of the economy. These support come from groups of enterprises, chambers, training centres, and other institutions. These initiatives are concerned with the slow process of technological innovation that SMEs have been experiencing and concentrate all the efforts toward the solution of this problem. Table 4.4 summarizes the network of institutions, by area of interest, currently working in SMEs sector.

3   Survey and characteristics of firms

3.1   Sample survey

As a first step, addresses of firms were taken from the directories of several institutions such as the Ministry of Trade, the Chamber of Industries, the Chamber of the Alimentary Industries and the Ministry of Tourism.[9] To delimit an SME, the study relies exclusively on the number of employees, i.e., firms with 10–50 employees were consider small and firms with 51–100 employees were taken as medium-sized enterprise.

---

9 Strictly speaking, tourism is not a manufacturing sector. However, given the importance for the Costa Rican economy (it generates much more foreign exchange than any other sector) few firms are included in the analysis.

Telephone-based interviews were conducted in Instituto Latino Americano De Politicas Publicas (ILAPP) headquarters in San José, Costa Rica. In many cases we visited firms to collect financial data. At the end, 68 firms were surveyed, although the number of calls was much higher. Selection of the firms was made using two criteria: willingness to participate in the survey and location. SMEs in Costa Rica tend to concentrate in the Central Valley, a region where 60 per cent of the population and the core of the economic activity are concentrated. The study followed the same pattern as a result of which most of the firms belong to the capital.

Data collected from the survey presents mixed features. Information on ICTs' use was responded by practically all firms without any objection. More quantitative information, such as the one on accounting, financing and investment were collected only partially. The survey was conducted during July 2004 and February 2005. However, 81 per cent of the firms surveyed were located in urban zones and 19 per cent in rural regions.

3.2 General overview of the firms

3.2.1 Firm's history

The interviewing process displayed a wide variety of situations regarding ownership of the firms. Most of the firms (42.6 per cent) are *Anonymous Societies*, although some of them were originally family business. The second largest group of firms, i.e. 27.9 per cent was owned by one person. Percentage of inherited family business was 14.7 per cent and the same percentage of firms was partly-owned by General Managers (GMs).

3.2.2 Production characteristics

Of the 68 firms interviewed, 86.7 per cent respond to manufacturing-related activities, 5.9 per cent to construction businesses and the remaining 7.4 per cent to tourism. Among manufacturing firms 27.9 per cent of firms belonged to Food and Beverages while Rubber and Plastic products manu-facturing firms constituted 14.7 per cent of the sample. Out of the remain-ing manufacturing firms, 13.2 per cent were engaged in manufacturing of Chemicals and Chemical products whereas the rest of the firms, i.e. 30.9 per cent were in the business of basic metal, paper products, machinery and equipment, and textiles production.

The sample of firms presents a wide range of situations regarding change in product mix over the last 10 years. Twenty-six per cent of firms did not change the product profile while the same percentages have experienced major changes in product mix. The remaining firms made moderate changes in their product-mix. Some firms, particularly in tourism, experi-enced an accelerated growth during the last 10 years that forced them to improve their infrastructure, the quality of their staff and the portfolio of their supplementary services. Others, like paper-related industries, observed the introduction of new and improved design software in the country.

Finally, several others still rely on effective traditional manufacturing methods, textiles being one of the examples.

3.2.3   Financial and economic performance

Of the 68 SMEs interviewed, only 29.4 per cent provided financial information. The low rate of response was a result of SMEs reluctant attitude towards making this sort of information public. Results gathered from SMEs willing to provide this data are presented in Appendix Table 4.1. Average sales turnover was Colon 710 million while on an average roughly 44 per cent of output was sold in international markets.

3.2.4   Use of Information and Communication Technology

A great amount of possibilities come into play with the inclusion of ICTs in production and non-production processes. The adoption of ICTs provided opportunities for making communication and transaction costs cheaper, production process more operational, and for introducing international standards in plants and marketing and for simplifying management.

All the firms involved in the study stated that at least one kind of ICT that was used by all was fixed telephones. The rate of use of ICTs, however, changed significantly with the complexity of ICTs. As shown in Figure 4.7, usage of email was 92.6 per cent, contrasting with the low rate presented among Computer-Aided-Design/Computer-Aided-Manufacturing (CAD/CAM) and Computer-Aided-Engineering (CAE) technologies, with only 4.4 per cent

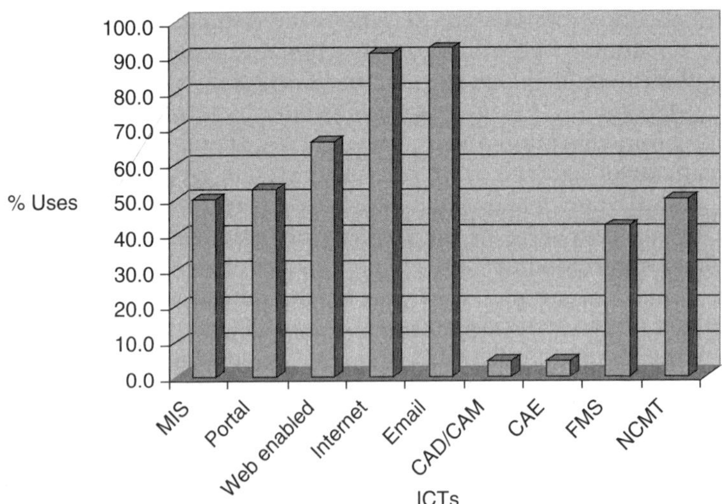

Figure 4.7   Degree of ICT adoption

usage rate. More than 90 per cent of the firms used Internet while adoption of portal-based technologies was 52.9 per cent.

Regarding other tools, only 50 per cent of the firms had a web page – and that too merely informative tool – because systems of virtual access to the services have not become functional yet in the country. Information systems, such as management information systems (MIS), are also installed in 50 per cent of the firms. Most firms used computers interconnected trough Local Area Networks (LANs) provided by Microsoft or by post switches which then made possible for all staff in the firm to have access to the public information. Till date, the country did not have adequate technology for exploiting advanced Internet services, limiting the capacity of SMEs to use them for trading products. Very often, problems related to this deficient infrastructure can cost some firms losses because it restricts the possibility of selling in other markets.

Two technologies, flexible manufacturing systems (FMS) and computerized numerically controlled machine tools (CNC on NCMT) are present in less than half of the firms. Although most processes were not 100 per cent automatic, several firms reported levels of automation at 80 per cent. It was reported by firms that computerized numerically controlled machines tools were only partially used in the production process and not through the whole production processes. Regarding FMS, the story is similar, and most SMEs do not have proper incentives to use such systems in their production processes.

3.3.5 Age of the firm and ICTs usage

A significant number of firms have been recently founded, as shown in Table 4.5. Accordingly, 38.2 per cent of the companies were founded after 1990 while only 25 per cent were established before 1974. In general, the SMEs in Costa Rica are relatively young firms. Age structures of firms and managers suggest that SMEs have the potential to easily adapt to new technologies, if they are available. Firms that started out this decade or the previous decade have an advantage with the introduction of ICTs in any of the added value chain or managing activities (by far the main field where ICTs can make most of their contributions). Older firms have the advantage of stable markets that have sustained them over time, and the capacity to withstand crisis and flexibility to adapt to changes and new technologies.

It was expected that older firms would have been less innovative and dynamic towards technological improvements and high-tech production processes as SMEs do not have enough resources at their disposal to experiment with new technologies. They adopt technologies only when the returns are ensured and gestation period is not very long. According to that logic firms should have been reluctant to introduce ICTs such as CNC, and FMS. However, results presented in Table 4.5 suggest that some older firms introduced technological changes in their productive, management and marketing activities.

Table 4.5 ICTs adoption and age of firms

ICTs	Year of establishment			
	Before 1974	**1974–1990**	**1991–1995**	**1996+**
MIS	7 (20.6)	13 (38.2)	5 (14.7)	9 (26.5)
Portal	9 (25.0)	15 (41.7)	7 (19.4)	5 (13.9)
Web enabled	11 (24.4)	19 (42.2)	9 (20.0)	6 (13.3)
Internet	14 (22.6)	23 (37.1)	13 (21.0)	12 (19.4)
Email	15 (23.8)	23 (36.5)	13 (20.6)	12 (19.0)
CAD/CAM	1 (33.3)	1 (33.3)		1 (33.3)
CAE	1 (33.3)	1 (33.3)		1 (33.3)
FMS	4 (13.8)	11 (37.9)	10 (34.5)	4 (13.8)
CNC	8 (23.5)	11 (32.4)	8 (23.5)	7 (20.6)
Total	17 (25.0)	25 (36.8)	14 (20.6)	12 (17.6)

Note: Figures in parentheses are row percentage.

In fact, the intensity of ICT adoption in older firms (established before 1990) was more than the new comers. This is true irrespective of the ICT tools adopted by the sample firms. For instance, 66.6 per cent of web using firms came into existence before 1990. This trend continues almost in case of every type of ICT using firm. However, the gap declined in advanced ICTs such as FMS and CNC using firms. Nearly 52 per cent of FMS using firms were established before 1990 while the percentage of CNC using firms that existed before 1990 was 55.9 per cent. The results suggest that experience contributed more than the entrepreneurship of GMs in the adoption of ICTs.

3.2.6 Managers' characteristics

General Managers were mostly males (96 per cent) and sometimes the manager was also the owner of the firm. On the other hand, 3 per cent of the general managers were females, but in none of the cases they were owners of the company. The average age of GMs was 46 years, although, individually, 50 per cent of GMs were less than 40 years of age. While 32.4 per cent of GMS were between 50–59 years old, the percentage of GMs older than 60 years was 8.8 per cent. Remaining GMs, i.e. 8.8 per cent preferred not to disclose their age. Two of the firms were just plants operating in the country as branches of the parent company. Hence we do not have information about their GMs' age, sex, and qualifications. In terms of educational background, 19.1 per cent of GMs did not have university degree whereas 8.8 per cent were graduates. The percentage of GMs with engineering degree was 22.1 per cent while 27.9 per cent were holding MBA and Ph.D degrees. Roughly 22 per cent of GMs did not disclose their academic background.

Manager's qualifications may have significant impact on ICTs use in Costa Rica. Although success of ICTs depends on the knowledge and training received by its users, its acquisition is significantly influenced by the knowledge base of GMs. The vision of head of the firm influenced technological improvement in the firm.

Table 4.6 presents the distribution of firms by the academic background of GMs and the adoption of ICTs. Academic qualification has been grouped into four categories, namely under graduates, gradates and post graduates, GMs holding engineering degrees, and finally professional degrees such as MBA holders GMs. GMs of 15 firms did not disclose their qualification. Hence, the analysis is based on 53 responses only.

Table 4.6 shows that the percentage of BE or professional degree holding GMs was highest in FMS using firms. Flexible manufacturing systems are considered advanced and costly ICT tools and the adoption of such tools requires thorough knowledge of the technology. Usefulness of FMS systems can be properly assessed by a qualified person. Consequently, most of the FMS using firms are managed by highly qualified staff. On the other hand, the adoption of tools such as CAD/CAM or CAE does not necessitate a highly qualified manager to understand its feature. They are usually purchased as a package and do not require customization. This is reflected in the case of sample firms also. The results illustrate a trend among SMEs in Costa Rica, showing that firms managed by general managers with higher levels of education used more advanced ICTs.

Table 4.6   Academic background of GMs

ICTs	GMs Education			
	Under graduate	Graduate Post graduate	Bachelor of Engineering (BE)	Professional degree
MIS	5 (18.5)	2 (7.4)	8 (29.6)	12 (44.4)
Portal	3 (10.3)	3 (10.3)	11 (37.9)	12 (41.4)
Web enabled	5 (14.3)	4 (11.4)	11 (31.4)	15 (42.9)
Internet	10 (20.4)	5 (10.2)	15 (30.6)	19 (38.8)
Email	10 (20.0)	6 (12.0)	15 (30.0)	19 (38.0)
CAD/CAM	1 (50.0)		1 (50.0)	
CAE	1 (50.0)		1 (50.0)	
FMS	2 (8.3)	2 (8.3)	9 (37.5)	11 (45.8)
CNC	4 (13.3)	2 (6.7)	10 (33.3)	14 (46.7)
Total	13 (19.1)	6 (8.8)	15 (22.1)	19 (27.9)

Note: Figures in parentheses are row percentages. Percentages are unlikely to add to 100 per cent as the missing cases are not reported.

3.2.7   Workers' profile and ICTs usage

The average number of workers per SME in the sample was 33 employees. The percentage of firms with less than 15 workers was 17.6 per cent while the percentage of firms employing between 15 to 24 workers was 27.9 per cent. Nearly 27 per cent of firms employed more than 50 workers.

As per the theoretical approach on scale economies, as a firm experiences growth and becomes larger, its need for ICT usage increases. In this study the number of workers has been used as a proxy of size as most of the firms declined to share financial data with us. The scenario of sample SMEs according to size and use of ICTs is presented in Table 4.7. It can be seen from Table 4.7 that largest number of MIS using firms, i.e., 32.3 per cent employed more than 50 workers. This is not only true for MIS using firms but for any type of ICT using firms. For instance 29 per cent of Internet using firms employed more than 50 workers while the percentage of such firms employing less than 15 workers was 19.4 per cent. The results presented in Table 4.7 suggest that size is important for the adoption of any type of ICTs.

The results presented in Table 4.7 show that SMEs need to adopt ICTs as they become larger with the purpose of lowering transaction costs and improve operation and management activities. The sample showed that smaller firms tend to use less sophisticated ICTs. For instance 25.0 per cent of firms with employment size of less than 15 workers were using portal-based ICTs while percentage of such firms employing more than 50 workers was 66.7 per cent.

Table 4.7   ICTs adoption and the size of firms

ICTs	Size of employment				Total
	< 15	15–24	25–49	50+	
MIS	7 (20.6)	7 (20.6)	9 (26.5)	11 (32.4)	34 [50.0]
Portal	3 (8.3)	10 (27.8)	11 (30.6)	12 (33.3)	36 [52.9]
Web enabled	8 (17.8)	12 (26.7)	11 (24.4)	14 (31.1)	45 [66.2]
Internet	12 (19.4)	16 (25.8)	16 (25.8)	18 (29.0)	62 [91.2]
Email	12 (19.0)	17 (27.0)	16 (25.4)	18 (28.6)	63 [92.6]
CAD/CAM			1 (33.3)	2 (66.7)	3 [4.4]
CAE			1 (33.3)	2 (66.7)	3 [4.4]
FMS	4 (13.8)	8 (27.6)	8 (27.6)	9 (31.0)	29 [42.6]
CNC	3 (8.8)	11 (32.4)	9 (26.5)	11 (32.4)	34 [50.0]
Total	12 (17.6)	19 (27.9)	19 (27.9)	18 (26.5)	68

Note: Figures in parentheses are row percentages while in square brackets are the percentages of firms that adopted a particular ICT.

3.2.8   Skill intensity and ICTs

The distribution of firms by use of ICTs and skill intensity of workers is presented in Table 4.8. The average share of workers with engineering degree was found to be (3.26 per cent). This is quite expected as workers with this qualification usually perform managerial functions while graduate and post-graduate workers are employed in non-production processes such as accounting and office management. Although diploma holders and technically trained workers are employed in production processes, their percentage is usually low as they are not for routine jobs. Usually persons with no formal training are employed in production processes for routine activities. Hence, their percentage is higher compared to other workers. Skill intensity of graduate/post graduate workers was 3.45 per cent while 79.26 per cent of workers did not have any formal training. The percentage of technically trained

Table 4.8   Skill intensity of workers and the adoption of ICTs

ICTs	Number of workers					
	Engineers	Graduate/ Post graduate	Diploma holders	Under graduate	Technical training	Other workers
MIS	46 (3.76)	39 (3.19)	70 (5.72)	77 (6.29)	21 (1.72)	1224 (79.33)
Portal	53 (3.87)	50 (3.65)	73 (5.33)	83 (6.06)	81 (2.26)	1369 (78.82)
Web enabled	57 (3.58)	55 (3.45)	87 (5.46)	131 (8.22)	31 (1.95)	1593 (77.34)
Internet	71 (3.38)	70 (3.33)	102 (4.85)	153 (7.28)	43 (2.04)	2103 (79.13)
Email	71 (3.35)	70 (3.31)	102 (4.82)	153 (7.22)	43 (2.03)	2118 (79.27)
CAD/CAM	6 (4.29)	10 (7.14)	11 (7.86)	32 (22.86)	1 (0.71)	140 (57.14)
CAE	6 (4.29)	10 (7.14)	11 (7.86)	32 (22.86)	1 (0.71)	140 (57.14)
FMS	35 (3.43)	46 (4.51)	51 (5.00)	63 (6.17)	24 (2.35)	1021 (78.55)
CNC	39 (3.27)	42 (3.52)	46 (3.86)	60 (5.03)	19 (1.59)	1192 (82.72)
Total	71 (3.26)	75 (3.45)	93 (4.28)	168 (7.72)	44 (2.02)	2175 (79.26)

Note: Figures in parentheses are row percentage.

workers was merely 2.02 per cent whereas percentages of diploma holders and undergraduate workers were 4.28 per cent and 7.27 per cent respectively.

It can be seen from Table 4.8 that two categories of skills, namely persons with engineering degree and technically trained persons have been kept separate. This is because technically trained persons were those who had a very specific training, and the duration of training was usually 6 months to 1 year whereas engineering graduates with 4/5 years training were employed for managerial functions and for very complex tasks in the production processes. Technically trained persons were usually employed in production processes while diploma holders were employed at supervisory level and in marketing activities.

Results show that in general technically trained persons were less than any category of workers. However, in case of CAD/CAM and CAE using firms the percentage of engineering graduates was highest, i.e. 4.29 per cent while such firms employed only 0.71 per cent of technically trained persons. There were only three firms that were using CAD/CAM and CAE. They were basically engaged in design activities. They did not require specially trained people to do the job. Table 4.8 also shows that simplest ICT tool, i.e., email using firms employed 3.35 per cent of engineers while percentage of engineers in portal using firms was 3.87. The results suggest that there is variation in skill intensity in different type of ICT using firms. Statistical significance of variation will be estimated in the subsequent analysis.

3.2.9 International partnerships/collaboration

In general, internationalization of firms in Costa Rica has been very limited. This is reflected by the fact that only 19.1 per cent of sample SMEs had inter-

Table 4.9   International orientation and the use of ICTs

ICTs	Foreign collaboration	
	Yes	No
MIS	10 (29.4) [76.92]	24 (70.6)
Portal	8 (22.2) [61.54]	28 (77.8)
Web enabled	11 (24.4) [84.62]	34 (75.6)
Internet	13 (21.0) [100.0]	49 (79.0)
Email	13 (20.6) [100.0]	50 (79.4)
CAD/CAM	1 (33.3) [7.69]	2 (66.7)
CAE	1 (33.3) [7.69]	2 (66.7)
FMS	7 (24.1) [53.85]	22 (75.9)
CNC	7 (20.6) [53.85]	27 (79.4)
Total	13 (19.1) [100.0]	55 (80.9)

Note: Figures in parentheses are row percentages while in square brackets are column percentages.

national partners of some kind or the other (Table 4.9). The low share of firms with this type of agreements can be the result of either traditional idea of keeping the business within the family or due to the difficulties in getting adequate technology for matching the conditions of firms with international partners. Majority of foreign collaborating firms (53.8 per cent) had collaboration for importing machinery, and drawing and design.

Table 4.9 also shows that the incidence of foreign collaboration was highest among email and Internet using firms. For instance, all the firms that had foreign collaboration were using email and the Internet. On the other hand 53.85 per cent of CNC or FMS using firms had collaborated with a foreign firm. Merely 7.69 per cent of internationally-oriented firms were users of CAD/CAM or CAE. The results suggest that one out of three CAD/CAM using firm was doing business in the international market while the other two were dealing in the domestic market.

3.2.10   Benefits of foreign collaboration

GMs were asked to express their opinion about benefits of foreign collaboration on a five-point scale ranging 1 for 'not important' to 5 'most important'. Average responses are depicted in Figure 4.8. It can be seen from Figure 4.8 that 'Good Will' has been rated as the major benefit of foreign collaboration. On the other hand, factors such as technology, brand name, and international linkages qualified with average. Opinion of GMs was also sought on role of ICTs in foreign collaboration. Roughly 62 per cent of firms reported a very strong role of ICTs in foreign collaboration.

3.2.11   Reasons for using ICTs

The main reason for introducing ICTs in SMEs was innovation within the firm. Twenty-five per cent of the firms stated that they introduced ICTs to face internal competition, while 29.4 per cent said that it was for facing

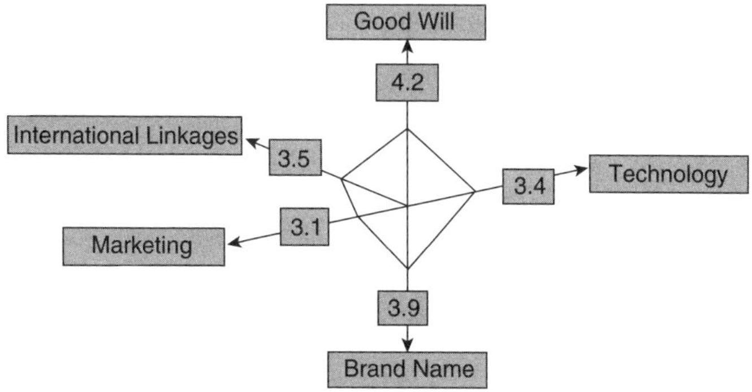

Figure 4.8   Benefits of foreign collaboration

external competition. Within the firm, management and production efficiency became core issues addressed by managers on why to use ICTs. In general, 38.2 per cent responded that ICTs were strongly needed for managerial functions 33.8 per cent for production efficiency.

Improved perspectives are other detected advantage provided by ICTs, as they allowed companies to track market trends. When asked if adoption of ICTs was a key factor in developing market monitor systems, 27.9 per cent answered that ICTs permitted tracking market information. However, implications in the production process for ICTs were not that strong because only 30.9 per cent reported that they did not experience any design advantages and only 22.1 per cent stated that ICTs were strong tools of design activities. External competition was attributed as the second most important reason to use ICTs. This could be because of the fact that many firms attempting to enter into international market must be globally competitive. These results suggest that ICTs were greatly used in management and marketing activities and not so much in the production processes. Although production in SMEs is highly mechanized, these automatic tools can only be explained in some areas, not being a recurrent characteristic of Costa Rican SMEs.

We sought opinion of GMs about importance of various reasons for adopting ICTs. It was reported on a five-point scale ranging 1 as 'not important' and 5 as 'most important'. The average scores of their opinions are presented in Table 4.10.

It can be seen from Table 4.10 that GMs of MIS and Portal using firms assigned highest rank to 'better management control'. It is quite obvious because the main contribution of MIS is to facilitate in corporate management. Second most important reason for the adoption of ICTs cited by MIS

Table 4.10   Reasons for using ICTs

ICTs	Average score				
	Market trend information	Internal competition	External competition	Efficiency in production	Management control
MIS	4.03	3.24	3.65	3.97	4.15
Portal	3.77	3.20	3.77	4.06	4.17
Web enabled	3.84	3.30	3.75	3.53	3.96
Internet	3.68	3.39	3.73	3.72	3.97
Email	3.64	3.35	3.73	3.67	3.92
CAD/CAM	4.00	1.33	1.67	3.67	3.33
CAE	4.00	1.33	1.67	3.67	3.33
FMS	3.93	3.38	3.88	3.93	3.90
CNC	3.79	3.53	3.93	3.73	3.97
Total	3.58	3.29	3.63	3.59	3.79

using firms was the ability of ICTs in providing market trend information. Understandably in the era of globalization market trend plays a crucial role in the performance of firms. Market trend does not necessarily relate to final products but might be related to inputs as well as production technologies. Overall almost similar weight was given to efficiency in production caused due to the adoption of ICTs. Table 4.10 also shows that overall (last row) external competition was cited as the second most important reason for the adoption of ICTs. This is reflection of the entry of MNCs in the country as well as the competition faced by export-oriented sample SMEs in international markets.

3.2.12   ICTs access constraints

In general access to ICTs in Costa Rica is not discriminative. Most of the firms stated that they did not have any problems accessing the net. This is because the National Institute of Electricity (the only Internet Service Provider in Costa Rica) establishes an open-access policy to all those firms requesting the service. Although this policy may seem very useful, it has its limitations because the organization works as a State monopoly that provides only one kind of technology, which not always turns out to be the best for entrepreneurs.

In general, most of the firms considered that Internet services were accessible in Costa Rica but only for general purposes, not for special tasks. Also, regarding cost of communication and Internet subscription, nearly half of the sample considered it also reasonable. Finally, firms positively evaluated the availability of trained human resources, something considered as an outcome of the national policy to enhance the capabilities of the labour force. Roughly 49 per cent of firms reported that lack of infrastructure was also not a constraint.

The average opinion of GMs recorded on a five-point scale ranging from 1 as 'not a constraint' to 5 as 'severe constraint' is presented in Table 4.11. It can be seen from Table 4.11 that average score of any potential constraints is less that score '3' which is labeled as normal constraint while seeking the opinion of GMs. Among the average score of constraints presented in Table 4.11, 'Internet speed' and 'communication speed' have been given more or less similar weight. Availability of Internet access points has been given the lowest rank, i.e. 2.5.

3.2.13   Consequences of using ICTs

In some industries ICTs became a source of competitiveness, while in others they only reinforced an already existing competitive advantage. In the vast majority, ICTs have only represented an additional tool for management. Productivity gains and deskilling were qualified as average, mostly because production and management process are already structured within SMEs, and in most cases they only came to reinforce and feed the established structures. A majority of firms (85.2 per cent) reported that use

Table 4.11 Constraints in using ICTs

ICTs	Average score			
	Internet access	Communication speed	Internet speed	Usefulness of information
MIS	2.35	2.55	2.68	2.52
Portal	2.29	2.56	2.54	2.60
Web enabled	2.58	2.82	2.87	2.61
Internet	2.55	2.84	2.84	2.73
Email	2.52	2.81	2.81	2.70
CAD/CAM	1.00	3.00	1.50	1.00
CAE	1.00	3.00	1.50	1.00
FMS	2.46	2.83	3.04	2.93
CNC	2.21	2.79	2.82	2.82
Total	2.50	2.75	2.78	2.67

of ICTs was either neutral to employment change or leads to an increase while a meagre percentage (2.9 per cent) were of the opinion that adoption of ICTs results in job reduction. Remaining firms (11.8 per cent) were not sure about the impact of ICT use on employment.

The reduction in transaction costs with the introduction of Internet and fibre optics has been one of the key positive consequences of using ICTs in SMEs, specially associated with lead time advantages, which were the most highly scored items. Lead time advantage was reported as average to strong benefit by 64.8 per cent of firms. The decrease of communication costs as a result of the introduction of virtual phones conversations and messenger programmes as well as the information accessibility caused by the ICTs, represented an incredible display of time reduction in marketing and distribution process, thus creating a time advantage in most enterprises. Productivity gains, flexibility in product design, and deskilling were other benefits reported by firms.

Opinions of GMs on the consequences of the adoption of ICTs were measured on a five-point scale ranging from 1 as 'no benefit' to 5 as 'strong'. The average scores are presented in Table 4.12.

Table 4.12 shows that 'Lead time advantage' has been given highest rank by GMs while deskilling and productivity gains were assigned same weight. Contributions of ICTs in modular and flexible design products have been assigned lowest rank. This could be because of limited number of sample firms were using ICT tools that allow flexibility in product designs.

3.2.14 Sources of competitiveness

The firm's capability to introduce technological improvements and innovations is a deciding factor on SMEs way of doing business. Constant

Table 4.12   Benefits of ICT use

ICTs	Average score			
	Productivity gains	Deskilling	Lead time advantages	Flexibility in product design
MIS	3.29	3.53	3.85	3.20
Portal	3.28	3.55	3.81	3.12
Web enabled	3.29	3.29	3.73	2.87
Internet	3.19	3.25	3.80	3.04
Email	3.16	3.18	3.76	3.00
CAD/CAM	2.33	1.00	3.67	3.33
CAE	2.33	1.00	3.67	3.33
FMS	3.03	3.22	3.69	3.22
CNC	3.03	3.30	3.71	3.00
Total	3.14	3.14	3.75	3.00

improvement, the generation of technologies and strategic alliances were perceived as turning points in Costa Rican SMEs. Nearly 52 per cent of the SMEs interviewed confirmed that they strongly operated using a product quality-approach though such an index maybe biased by the firm's public need to defend their products. Product design was reported as strong source of competitiveness by 35 per cent of the firms interviewed showing the firms' ability to adapt and diversify. Advertisement on the other hand, was overlooked by SMEs. R&D was not qualified as a strong competitive factor because most SMEs cannot base their production on investigation processes. Costa Rican SMEs constantly struggle in the context of globalization and expanding markets and the need to rely on short-term profits does not leave room for long-term plans such as R&D.

Market share or size of the operation, was reported by most firms as average because most SMEs cannot monopolize the entire market. Despite most SMEs had limited market participation they had already endured a long time in the market. Delivery schedule was considered as a strong competitiveness source by 48.5 per cent Costa Rican SMEs analysed. This responds to a number of factors present on Costa Rica's economy such as closeness to the oceans on both sides of the country and the ability for small firms to quickly adapt to customers needs. On the other hand, 58.8 per cent of SMEs rated brand name as sources of competitiveness from average to strong. A correlation between product quality and brand name may exist because of the market situation it implies. Nearly 40 per cent of the firms stated that they did not compete on the basis of low salaries.

Like consequences of ICT adoption, sources of competitiveness were also measured on a five-point scale. Average scores are presented in Table 4.13. The table shows that product quality was rated as a very strong source of

Table 4.13   Sources of competiteoness

ICTs	Average score			
	Product quality	**Flexibility in design**	**Delivery schedule**	**Band name/ goodwill**
MIS	4.53	3.20	4.21	3.71
Portal	4.53	3.12	4.30	3.94
Web enabled	4.50	2.87	4.30	3.51
Internet	4.48	3.04	4.36	3.58
Email	4.45	3.00	4.37	3.60
CAD/CAM	5.00	3.33	5.00	1.67
CAE	5.00	3.33	5.00	1.67
FMS	4.59	3.22	4.38	3.79
CNC	4.59	3.00	4.45	3.64
Total	4.45	3.64	4.39	3.59

competitiveness followed by delivery schedule that was rated as 4.39. The other sources that were fairly ranked were flexibility in product design and brand name/good will.

3.2.15   Human resource development and training

The quality of human resource and the training they undergo play a key factor increasing SMEs capabilities. Human resource is a rather dynamic force that needs constant innovation. Therefore most SMEs implement programmes to put their staff up to date. Analysis of the data collected from survey suggests that 43 per cent of firms reported that between 76 and 100 per cent of their workforce received some kind of training in the last 5 years. In Costa Rica there are two existing sources of learning, namely: on-job training, and outside SMEs premises. There are two institutions that provide external training. One is a public institution called INA (by its Spanish initials the Learning National Institute) and the other is through chambers, organizations and programmes facilitated by the Costa Rican Ministry of Economy.

The Learning National Institute provides short courses on basic industrial practices and mechanic activities. According to some SMEs interviewed (although this institution was not included in the questionnaire, many firms during the interview complained) this institution could improve a lot more. One of the many critics made on this institute is the short duration of courses and the simplistic approach of their training.

Chambers and the Ministry of Economy were less critiqued because they developed training concentrated on the chain of command and therefore it has a higher degree of insightfulness. One good example would be the chamber of Alimentary Industry, which directs its efforts at reinforcing sanitary measures, standards and norms related to production procedures.

Table 4.14 Human resource development and training

Learning processes	Degree of effectiveness				
	Not effective	Poor	Average	Good	Very good
Formal training	5.9	4.4	27.9	25.0	17.6
Training by other institutions	8.8	2.9	13.2	27.9	23.5
Learning by doing	2.9	1.5	10.3	33.8	35.3
Searching the internet	17.6	10.3	17.6	13.2	10.3
Learning from technical partners	11.8	5.9	11.8	14.7	7.4
Overseas training	16.2	4.4	5.9	13.2	10.3

Keeping in mind the role of training in the acquisition of knowledge, we sought opinion of GMs on effectiveness of learning processes. Importance assigned by GMs to various modes of trainings is presented in Table 4.14. Largest per cent of firms (35.3 per cent) rated 'Learning by Doing' as most effective way of training, while 23.5 per cent of firms ranked training by other institutions as the second best mode of learning.

Formal training was considered the third best mode of skill up-gradation. The results show that searching the Internet, overseas training, and learning from technical partners were considered effective with only 10 and 7 per cent of firms indicating that this kind of training is highly useful. This situation can be attributed to lack of financial support and strategic vertical and horizontal links. Many firms stated that training through these modes, especially with knowledge found on the Internet and obtained abroad presented application obstacles. Average scores are presented in Appendix Table 4.2.

4 Theoretical framework and statistical analysis

Having discussed the association of firms' characteristics with the adoption of various ICT tools, we identified factors that separated different levels of ICT using firms. We have categorized firms into three groups, namely low level of ICT using firms, moderate users of ICTs, and advanced ICT using firms. Cluster analysis has been used to group sample firms. ICTs such as email, Internet, web enabled technologies, portals, MIS, FMS, CNC, CAD/CAM, and CAE have been included in the cluster analysis. Before presenting and discussing statistical results, we present a theoretical framework (Fig. 4.9) and hypotheses tested in this study.

Hypothesis I: As depicted in Figure 4.9, we hypothesize that conduct of firms represented by skill of workforce, investment on R&D and innovative activities, quality consciousness, and international orientation influence

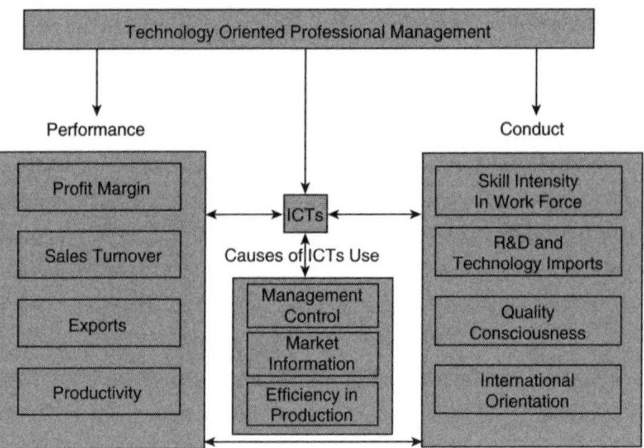

Figure 4.9   Theoretical framework

and are influenced by the adoption of ICTs. Skilled workforce is needed for not only using ICTs effectively in production processes but also in marketing, after sales support, and other coordination activities. Skill workers could be more effective in using technologies such as customer relation management and supply chain management (Mehta, 2000). Although R&D may not be a common phenomenon in SMEs, the import of new and efficient production technologies is expected to be higher in advanced ICT using firms. This might be due to the fact that the use of new production technologies requires extensive interaction with suppliers of such technologies that can be easily accessed through Internet and email. The import of new technologies might be associated with the adoption of ICT not only for their acquisition but also for searching appropriate technologies.

The adoption of ICTs might be more extensive in the firms that are quality conscious and export-oriented. In the era of globalization, the quality of products is not determined only by using high quality inputs but also by the modularity and flexibility in product designs. This is very relevant in manufacturing sector. Quality parameters in service sector are different. In service sector, quality is determined by ability, willingness of workers and management and service infrastructure. In both the sectors ICT can greatly contribute to better quality of products or services. Programmable equipment used in production processes enables firms to manufacture modular and flexible design products whereas availability of mobile Internet and email services attracts customers in a service sector like tourism. International orientation is facilitated to a large extent by ICTs. In manufacturing sector ICTs may be used for import-export activities and in service sector they might be useful for maintaining or expanding international clientele by using Internet, email, and web enabled technologies.

Hypothesis II: Degree of ICT adoption is expected to be significantly influenced by the perceived benefits of their use. Several benefits such as better management control, reduction in market and event uncertainties, provision of latest market information, efficiency in production process, reduced transaction costs, and availability of timely, accurate, authentic, and reliable information have been cited in the literature (Pohjola, 2001). However, the adoption of ICTs in SMEs largely depends on the perception of general managers about potential benefits of ICTs. The causal relationship between benefits and the adoption of ICTs is bidirectional. We argue that perception of general managers could influence the adoption and the actual benefits in turn can result in the adoption of more advanced ICTs.

Hypothesis III: Finally, the adoption of ICTs is expected to contribute to better performance of firms. As depicted in Figure 4.9 performance indicators could be sales turnover, increase in exports, and higher profit margins, which may be due to augmentation in productivity, introduction of new products, expansion of markets etc. In this case also there exist a bi-directional relationship between intensity of ICT use and performance of firms. Another factor, which is very crucial in differentiating advanced ICT using firms from the rest, is the qualifications and knowledge base of general managers who are generally owners also. Hence, it is hypothesized that academic background of general managers influenced conduct of firms and the degree of ICT adoption and that in turn results in better performance of firms.

4.1   Statistical analysis

We have not been able to test the discriminating power of all the factors depicted in Figure 4.9 in separating advanced ICT using firms from the rest due to the lack of information provided by firms. Based on the description of firms characteristics presented in Section 3, analysis of variance of several factors is presented in Table 4.15. As can be seen from Table 4.15 the possible discriminants are categorized into six groups, namely: entrepreneurship, international orientation, causes and consequences of ICT adoption, competitiveness, and human resource development factors. The description of the variables is shown in Appendix Table 4.3. Although use of general manager's academic background as a proxy of entrepreneurship has been debatable, we have used GMs educational background and skill intensity of workers as representative of entrepreneurship.

Table 4.15 presents mean value of the variables in three groups of firms, namely: low ICT using firms, moderate users of ICTs, and advanced ICT using firms. F-value and significance of means are also shown in the

Table 4.15   Analysis of variance of discriminants of ICT use

Variables	Intensity of ICT use				
	Low	Moderate	Advanced	F-value	Sig.
1. Entrepreneurship					
MD_EDU	2.58	2.35	3.35	3.655	0.033**
ENG	0.96	1.04	1.47	0.872	0.423
2. International orientation					
FORN_COLL	1.92	1.80	1.65	2.626	0.080*
3. Causes of ICTs use					
MGMNT_CTRL	3.19	3.92	4.35	4.695	0.012**
MAR_INFO	3.04	3.61	4.29	5.032	0.009***
EFFI_PROD	3.23	3.50	4.25	3.075	0.053*
4. Consequences					
LEAD_TIME	3.62	3.88	3.71	0.268	0.766
PROD_GAIN	2.65	3.52	3.12	2.687	0.076*
FLEX_DESIGN	2.79	3.05	3.38	0.737	0.483
5. Competitiveness					
PROD_QUAL	4.42	4.38	4.59	0.587	0.559
DEL_SCH	4.50	4.24	4.43	0.568	0.570
6. Human resource development					
LEARN	4.00	4.36	4.07	0.839	0.438

Note: ***→ 1 %; **→ 5 % and *→ 10 % level of significance.

table. Table 4.15 shows that qualification[10] of general managers differed significantly in three types of firms. In fact the average academic qualification of GMs of low and moderate ICT using firms is similar. But there is significant difference in average qualification of GMs of advanced ICT firms from the rest.

Although average number of engineering graduate workers increases as the intensity of ICT use, it does not differ significantly in three groups of firms. Mean value foreign collaboration, which was measured on a binary scale, differs at 10 per cent level of significance. Table 4.15 shows that better management control, ability of ICTs in providing reliable information on market trends, and the adoption of ICTs induced efficiencies in production processes were assigned more importance by advanced ICT using firms. The opinion of GMs on productivity gains due to adoption of

---

10 Academic qualification has been quantified as follows: vale 1 is assigned to undergraduate GMs, 2 to graduate and post graduate GMs, 3 to bachelor of engineering degree holders, and highest rank, i.e. 4 is assigned to professionals such as MBA degree holders.

ICTs also differs significantly. However, their opinion of other consequences of the adoption of ICTs did not differ significantly. Mean value of other factors such as quality consciousness, delivery schedule, and learning by doing did not follow any trend.

Subsequently data were analysed in a multivariate framework. Results are presented in Table 4.16. A forward stepwise discriminant analysis was preferred over other multivariate and PROBIT models. This is because discriminant analysis does not pre-assume causal relationship between the group identification variable and others. As depicted in the theoretical framework, it is assumed in the study that several factors reinforce causal relationship. Hence, discriminant analysis was considered as the most appropriate technique to be used in this study. The procedure begins by selecting the individual variable that provides the greatest univariate discrimination (in terms of groups mean difference of F). It then pairs the first variable with each of the remaining variables to find out the combination that produces the greatest discrimination. The variable that contributes to the best pair is selected. In the third step, the procedure goes on to combine the first two with each of the remaining variables to form triplets. The best triplet determines the third variable to be entered, and so on. It stops the procedure when groups mean difference F is less or equal to 1. Table 4.16 presents the summary of the stepwise procedure and the discriminants selected with their relative contribution to the discrimination.

There are some variations in results presented in Tables 4.15 and 4.16. For instance academic qualification of GMs, which was significant in univariate analysis does not remain so in multivariate analysis. This is only due to the degrees of freedom available in each analysis. Results presented in Table 4.15 are by and large based on 68 firms except MD_EDU which is based on 53 firms. Univariate analysis does not require a common sample size whereas multivariate analysis results are based on a common sample of 35 firms. This is the main reason for variation in results presented in Tables 4.15 and 4.16.

Skill intensity, which is represented by a number of engineers working in firms shows highest level of discrimination (1 per cent). Productivity gains and efficiency in production process did not emerge significant in multivariate analysis while they were significant, though at 10 per cent level, in univariate test. Based on the discriminant function shown in Table 4.16, its discriminatory power was estimated. The results are presented in Table 4.17.

Out of 15 low level of ICT using firms used in the discriminant analysis, discriminant function is able to classify 12 firms correctly resulting in 80 per cent power of classification. Out of nine moderate ICT using firms, the function is able to classify 88.9 per cent firms correctly whereas the classification power of the function is 100 per cent in advanced ICT using firms. The overall classification power is 88.6 per cent. Classification power of discriminant function is an indicator of appropriateness of model

Table 4.16   Summary table of discriminant analysis

Variables	Wilks Lambda	F-Statistics	Sig.
1. Entrepreneurship			
MD_EDU	0.919	1.414	0.258
ENG	0.727	5.996	0.006***
2. International Orientation			
FORN_COLL	0.765	4.921	0.014**
3. Causes of ICTs use			
MGMNT_CTRL	0.815	3.631	0.038**
MAR_INFO	0.851	2.808	0.075*
EFFI_PROD	0.927	1.254	0.299
4. Consequences			
LEAD_TIME	0.986	0.223	0.801
PROD_GAIN	0.950	0.834	0.443
FLEX_DESIGN	0.996	0.059	0.942
5. Competitiveness			
PROD_QUAL	0.922	1.351	0.273
DEL_SCH	0.943	0.973	0.389
6. Human Resource			
LEARN	0.995	0.086	0.918

Note: ***→ 1 %; **→ 5 %, and *→ 10 % level of significance.

Table 4.17   Classification results

Actual group	No. of firms	Predicted group membership		
		Low ICT Users	Moderate ICT users	Advanced ICT using firms
Low ICT Users	15	12 (80.0)	1 (6.7)	2 (13.3)
Moderate ICT users	9	1 (11.1)	8 (88.9)	
Advanced ICT using firms	11			11 (100.0)

Classification power of the discriminant function: 88.6 per cent

specification. Having attained 88.6 per cent classification power, the model is considered as well beyond the acceptable limits.

4.2   Discussion of results

From Table 4.15, MDs qualification emerged significant discriminant though the level of significance was at 5 per cent. The emergence of MDs academic qualification as a significant factor is not only in line with existing literature but also supports the hypotheses of the study (Lal, 1996; Earl, 1989). Since the sample was dominated by manufacturing firms, ICT tools can be used in both peripheral and core activities. This places a demand on the MDs to be aware of the intricacies of ICT tools so that the potential benefits are fully reaped. Moreover, GMs of SMEs are never in a position to adopt technologies whose benefits are not assured.

The relationship between skill intensity and the intensity of ICT used is in accordance with the hypothesis of the study. The findings also corroborate with other earlier studies (Doms *et al.*, 1997; Rada, 1982). In fact ICTs are regarded as skill biased technological change. Although several proxies of skill intensity such as experience of workers and wages paid to workers have been used in earlier studies, we have considered the number of engineering graduates. We could have used percentage of engineering graduates and ordinary graduates/postgraduates in the total workforce. But we have preferred to use only engineering graduates as skill intensity, which is more relevant for the use of ICTs. We are not arguing that engineers are needed to use email and the Internet but they are certainly needed for implementing the ever-changing technologies such as portal and web enabled technologies. And this might be the reason of capturing the role of skill intensity in discriminating advanced ICT using firms from the rest.

Several scholars (Stiglitz, 1989; Kiiski and Pohjola, 2002) have emphasized that ICTs play an important role in exchanging information, knowledge, and product designs between manufacturers and suppliers of technology. One of the major contributions of ICTs in the business environment is to facilitate better co-ordination of manufacturing activities. Portal and web enabled tools may be the best-suited technology to co-ordinate with foreign companies particularly. Emergence of technological collaboration as important discriminant is a case in point. The results support findings of earlier studies (Oyelaran-Oyeyinka and Lal, 2005; Pohjola, 2001). Technologically collaborating firms need a greater degree of interaction with the suppliers of technology than other firms. Interaction consists not only in knowing the specification of imported equipment but also in sharing intangible and embodied technological knowledge. Codified knowledge such as engineering drawing, detailed design, and so on could be exchanged with the use of advanced ICTs more effectively.

Emergence of MGMNT_CTRL as one of the important discriminants substantiates the findings of earlier studies (Lal, 1996; Mehta, 2000). The use of office automation technology such as MIS might have been providing better control of information to GMs. Hence, GMs of advanced ICTs using firms, who were also users of MIS, might have given more importance to better management control. Although technologies such as local area network (LAN) and wide area network (WAN) have not been included in the study, it was noticed during the survey that few firms were these networking technologies and were found to be very effective for managerial functions.

Computer integrated assembly lines are expected to be more efficient than traditional ones. The efficiency derives from the structure and self-fault detection mechanism built in at crucial stages of manufacturing. For instance, after inserting electronic components such as Large and Very Large Scale Integrated chips (LSI and VLSI) in the PCB used in an electronic product, the input and output parameters of the module are checked at the assembly stage itself. And if there is any discrepancy found between the expected and actual parameters, PCB does not proceed to the next stage of manufacturing processes. It is sent automatically to the fault correction stage of the production process. Although there were very few firms engaged in manufacturing of electronic goods, the chemical and chemical products, machinery and equipment, and textiles firms might have experienced efficiency in production processes. This can be achieved by ICTs such as FMS and CNC. Consequently GMs of firms using such tools might have attributed high importance to efficiency in production processes due to use of ICTs.

Internet facilitates in searching information to a great extent. Information could be related to product specifications or configuration of production technologies. With the increasing use of Internet and related technologies, industry associations have developed tendencies to provide such information to their clients through non-traditional means. The relevant information can be easily searched and downloaded by firms. Production technology manufacturing firms are increasingly using their web sites for advertising latest specification of new technologies. This information can again be easily accessed by users of those technologies. Latest trends in design and composition of textiles are very crucial information for firms engaged in this business. Without use of the Internet, access of this kind of information was virtually impossible. Hence GMs of innovative firms might have assigned due importance to ability of ICTs in providing market information.

Although we did not go into details of whether the use of ICTs contributes in labour or capital productivity, question was mainly meant for labour productivity. Findings of the study corroborate with post-1995 studies on ICTs and productivity. We are not aware of any study until mid-1990s that found evidence of productivity gains due to the use of ICTs. However, after mid-1990s there have been several studies to show that advanced ICTs yield in higher labour productivity (Brynjolfsson and Hitt, 1996; Pohjola, 2001). Productivity gains in non-production activities come from the exchange of information electronically. The information exchange can take place between workers and management, within management groups, between firm and other business partners. The productivity gains in production processes stem from the use of programmable equipment.

5   Case studies

5.1   Alimentos Don Mariano

Generally, most small firms look for isolated markets to exploit its full potential. Snack products generally aim at mass consumption markets, so it is very strange to see small enterprises entering this kind of markets. Thus, when one firm like Alimentos Don Mariano appears on stage, it deserves special attention from the analysts.

Now what makes a small firm such as Alimentos Don Mariano a successful producer of mass consumption products? This firm transforms a daily product like cassava into a star product that is currently exported to the European Union and Canada, with expansion plans to some Caribbean countries and the USA as well. The successful story of this firm goes hand-to-hand with some recent changes applied in this firm in terms of modernization of the production processes, the introduction of high-quality human resources and the application of ICTs as an instrument to improve performance. Although most of these changes are of a recent nature, the success of the business can be perceived by analysing not only accounting data but also the achievements of an enterprise that started out as an informal business.

5.1.1   Historical background

The owner of this small enterprise, Industrial Mechanic, Claudio Díaz Gei, acquired great experience in the snacks industry by working with large Costa Rican companies such as Tosty, Jaja, Jack's, Alimentos Cariari and Mejores Alimentos. In 1991, job opportunities and projects were lacking and due to urgent needs, Don Claudio (as he is known) decides to repair an old industrial frying machine for selling purposes. In that precise moment, Don Claudio realized the potential earnings in the snacks business and decided to start his own firm by frying and selling packed cassava. With this objective in mind, he bought 20 quintals of product and with the support of five employees he began producing the first type of snack, salted cassava that was sold in informal kiosks in the country.

Don Claudio's instincts were correct. The business began generating profits, so he decided to expand the operation and opened the first plant in Curridabat, where his old factory used to be located. The first 2 years of the business were characterized by an expansion of the firm as well as by the huge amount of solid industrial wastes generated by the frying process. Within this period of time the second product presentation was conceived, cassava with chili and lemon.

Toasted cassava with barbecue was the third presentation that came out in 1993 and along with this new product also appeared the first printed package. Precisely during this year, the firm became an anonymous society called Don Mariano S.A. By 1994 the firm perceives the need to rent a larger establishment, because until that moment the productive process was executed in two places, one located in Guayabo of Curridabat where the cassava was peeled and then cut off, and the second one in Curridabat downtown, where it was fried or toasted and then packed. The larger place located in Barrio Guayabo de la Concepción concentrated the whole operation and contributed greatly to improvements in efficiency. In 1996, the productive process was reinforced even more when an old industrial toaster machine was bought from Café Rey, a national coffee producer. Don Claudio, using his skills in industrial mechanics, adjusted the machine to the needs of the firm.

The idea of a new building started in 1997, but it was not until 1998 that its construction began. The new factory, however, was inaugurated in 2001 and fully utilized until 2002, when the processes of peeling and cutting were moved from Guayabo to San Diego de Tres Rios, in the province of Cartago, where the industry is currently located. Between 1998 and 2002, the factory purchased a truck for distributing its product (1998) and a new frying machine (2000).

5.1.2 Structural organization and production process

The organizational changes experienced by this firm have been highly influenced by the incorporation of information systems. In addition to the snack factory, the building hosts two more activities, an industrial mechanic repairs centre (FREMBUSA) and another factory for producing fried tortillas (Tortillas La Campesina). The repairs centre was establishing for solving the problems of assembling, functioning and adaptation of the new industrial machines or for repairing the existing ones.

The nature of the frying and toasting process requires minimum skills, so when the firm started to operate mostly family members were involved in the production of snacks. As the enterprise grew and expanded, the firm

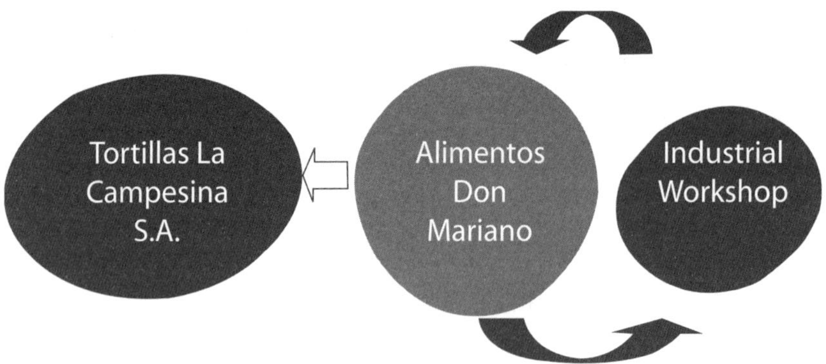

Figure 4.10 Alimentos Don Mariano group

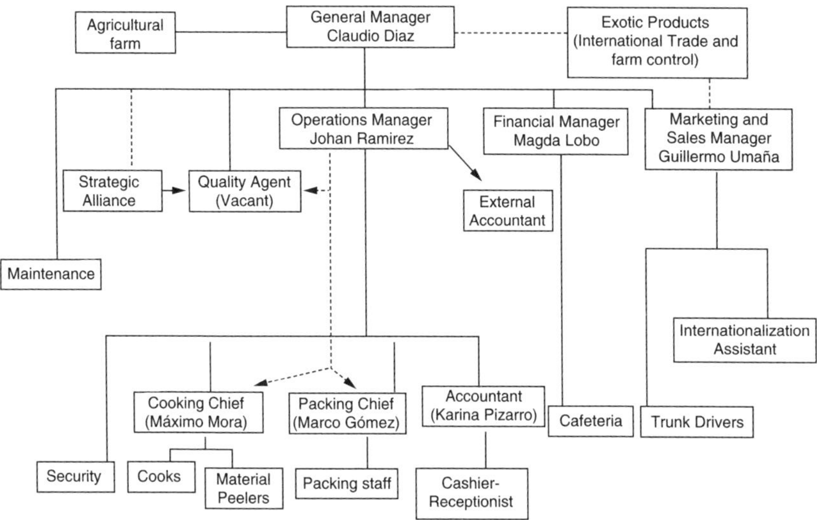

Figure 4.11   Organizational chart of Alimentos Don Mariano S.A.

saw the need to incorporate trained management staff. The new hired professionals made important contributions and substantially reorganize management, marketing and the overall strategy of the company. Part of those changes was the incorporation of ICTs. Figure 4.11 shows the new organizational of the firm by the end of 2004.

The production process takes place in one factory located in an industrial region north of San José. The firm is a family business founded on 1993 and has exactly 26 workers, although during the last years the number of employees has declined from 35 in 2001. Between 1991 and 2004, staff grew five times. The quality of the human resources stands out with two industrial engineers, two MBA, one food technician and all the rest with high school approved. That is, around one-fifth of Alimentos Don Mariano staff has a university degree. Precisely, the new information system was introduced by highly-educated persons in those areas shadowed in the previous chart.

The Production Manager, industrial engineer Johan Ramirez, reorganized the systems and put together the information issued by internal information systems, registration and documentation. On the other hand, Marketing and Sales Manager, MBA Guillermo Umaña, developed a strategy for increasing exports using email and Internet for communicating with potential clients and getting access to key information. That was the start of a new era for Alimentos Don Mariano because those two tools made possible the incursion in exporting, otherwise the high costs of visiting potential clients outside (with unclear results) would be a significant barrier almost impossible to overcome.

5.1.3 Survey results and ICTs innovation

The overall profile of the company shows that the firm is an average small company in many senses. Today the principal products for this enterprise are fried cassava, fried plantain and fried ñame.[11] When asked about the changes experienced by these products, the firm answered that the production considerably evolve during the last 10 years, both in quality and number of snacks produced. For instance, 10 years ago, the company only produced cassava, but now the firm diversified its lines of production to more than ten products and flavors. This has been done without any kind of external partnership.

Regarding the use of ICTs, the firm began using these technologies since the first year of operations, in 1991, when its first fix line was installed in the company. Later, in 1999 and 2000, some computers were introduced in the accounting department, and 1 year later, in 2001, additional computers were bought and interconnected with a 16 post switch as well as an ADSL system with a 64–128 kbps speed.

As for the technologies used, the firm makes extensive use of ICTs discussed in the chapter except CAD and CAE. The benefits achieved from using ICTs are mostly focused in the direct contact the firm can keep with foreign clients. When rating the importance of ICTs for the daily operations of the firm, the average score was four (out of five). Other positive consequences identified by the firm on the use of ICTs were productivity gains, deskilling, lead time advantage and flexibility in product designs.

The marketing process has evolved since the inclusion of several informatics improvements. The product was originally sold to informal vendors or peddlers and the target market was the public in general, mostly young people. The product was sold at schools and high-school, cafeterias and on the streets. As the business grew, the product was sold to wholesale traders, distributors, and outside the Central Valley. Most recently the introduction of metallic packages, new presentations and registration methods made possible the product marketing in supermarkets and international markets. The year 2004 marked the beginning of a new era for two reasons. One, the factory started exporting to international markets such as England and Canada and began contacting small customers in Hong Kong, USA and Mexico. Two, the relationships with domestic distributors changed, so now vendors go to the distributor and wholesale traders to purchase the product, and long-distance drivers come to the factory, pick up the merchandise and leave.

The level of competitiveness showed an enterprise that maintains its position in the national market due to quality product, design flexibility, a great deal of market share for being a lead firm, brand name, technological collabo-

---

11 A very popular tubercle of the Central and South American regions.

ration and local suppliers. In terms of human resources, despite the fact that no human resource department exists, nearly 80 per cent of the total of workers had received training of some kind. All training activities were qualified as good except for on-the-job training by other institutions and searching the Internet, which were qualified as excellent and average. Training has been concentrated on topics such as quality, product handling, and manufacturing practices, environment control and the HACCP, which refers to the standards related to the food industry. Average length per course is 1 month and the course is generally taken every 6 months. Internal training focuses on quality and inventory manipulation and lasted no more than 2 weeks.

Besides, in terms of infrastructure, all the facilities visited during the field trip showed adequate conditions for the elaboration of food products. The production process is 80 per cent automatic and the factory has a special sewage treatment with sediments and bacteria to prevent water pollution (see Fig. 4.12). Key constraints for continuous growth were the dearth of

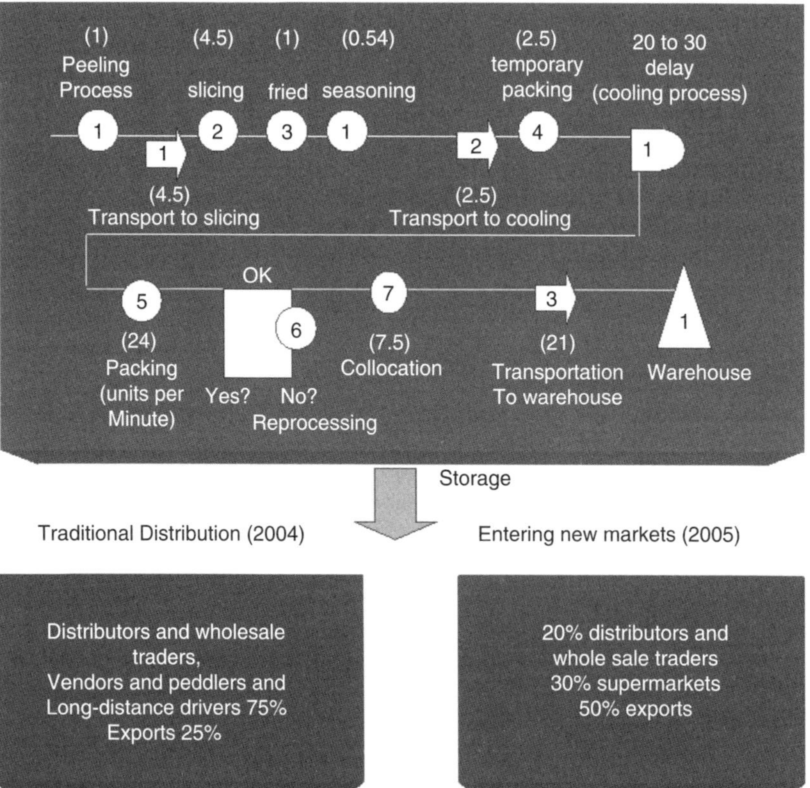

Figure 4.12   Alimentos Don Mariano product diagram

components and parts for the machines, the absence of foreign collaboration and the increased disloyal competition experienced in the domestic market, where product imitation by the informal sector is an everyday headache for the management. On average, those three factors were scored 3 out of 5 as the most important problems faced by the firm.

5.1.4 Economic and financial performance

Data on financial issues do not exist before 2001. The interview did not provide information related to assets, labour costs and sales turnover on a time series basis. Instead, all related information to economic and financial performance has to be calculated using the operation and overhead costs. The price of the product is the result of three components: 52 per cent raw material, 15 per cent labour cost and a 33 per cent cost that is the result of overhead plus sales turnover.

5.2 Hulera Costarricense

Firms established in their businesses for generations are excellent sources of how firms should be run and how to keep the company competitive despite time. The Hulera Costarricense, a 60-year old rubber factory, is one of those cases where time makes the firm functional and competitive.

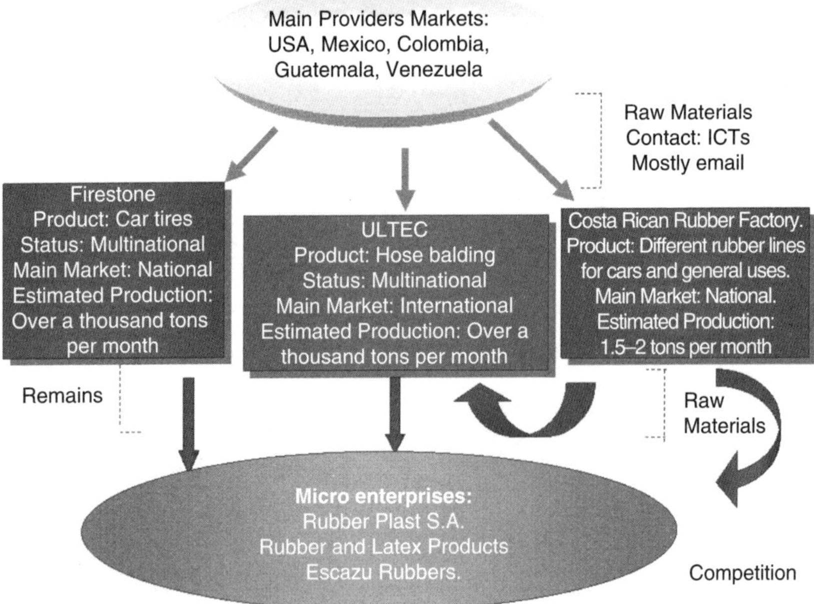

Figure 4.13 Costa Rica rubber market structure

The rubber market in Costa Rica is dominated by two large multinational firms (Fig. 4.13), ULTEC, dedicated to produce and export hoses baling around the world and Firestone, devoted to tires manufacturing for national and international markets. In total, the market has four additional firms, the Hulera Costarricense and three microfirms that catch the remaining market. Despite the existence of two multinational firms, the Hulera Costarricense is a leader in many lines of production. What are the factors that make this firm highly successful and what is the role of ICTs in promoting that success?

5.2.1   Historical background

Fernando Castro Avilés founded the firm in 1945 at the centre of San José. It started as a micro enterprise installed in a house and dedicated to the manufacturing of rubber-made women shoe heels, using a special tire-related product as raw material. Tires were obtained by scratching tires with motorized emery, which later took the adequate shape in the hands of the operator and a hand operated rubber press. In 1957, the production process diversified adding men shoe heels, plushy, doormats and car mats. The inclusion of the former products leads to an increase in the usage of raw materials.

By 1965, Rodolfo Borbón took office as General Manager of the company. In that year, the firm had ten employees and the first mill used to mix rubber began operations. The machine became a key event in the life of the firm because it enabled the elaboration of different mixtures of basic raw materials. Some years later, two bigger mills were bought and a mix press was acquired. Finally, the heating electric process changed to a vapour-based one which led to the first hydraulic rubber press.

By 1970, the firm became an anonymous society and started working as a new business. The shoe heels fabrication was left out the production process and the factory specialized in the elaboration of rubber mats for steps and buildings, which became the most important products for the firm. The fabrication of additional items by request also became an important sales generator. Today this firm is a medium enterprise located in Escazú (western part of the capital) with a plant of approximately 1,300 square meters and 32 employees in the production department and 16 employees in the sales and management department. The plant currently has an inner mixer or what is called a 'Bambury', three kneading, 12 hydraulic presses, three manual presses, one guillotine, two ovens and a fully accessorised laboratory that allows for measuring and functioning of materials used in the process. There is also a maintenance department. The raw materials used by this firm are imported from Guatemala, Mexico, Germany and the US. These materials are considered of the highest quality. The products are distributed in five different lines, namely: (1) individual mats for vehicles, (2) hanging drapery and mudguard for vehicles, (3) mating for steps, hallways, floors and sinks, (4) packaging for special pieces, and (5) industrial chemicals.

5.2.2   Firm profile and production lines

Regarding human resources, the firm mainly relies on industrial and chemical engineers for developing the production process and MBAs for managing. The core of the staff is characterized by being diploma holders and high school graduates.

The first years of the present decade have been characterized by a significant reduction in the number of employees mainly because the domestic market has been declining due to an increased competition. Between 1999 and 2005, the number of employees fell by 36 per cent. Also, according to the manager of the company, another triggering effect to the contraction of the payroll is the presence of sunk costs to keep a highly trained staff and a more advanced infrastructure. This firm competitiveness is based on quality product, flexibility design, market share and reputation of the 60-year old brand. When competition increased during the last few years, a significant share of the market was lost, decreasing competitiveness and leading to a reduction of the operations.

The production process of the company comprises five lines of production (Fig. 4.14), being two of them massive production processes (lines 2 and 4) and four of them specialized lines develop mostly to suit the clients' needs (2, 4, 5, 6). The production process consists in treating the rubber to

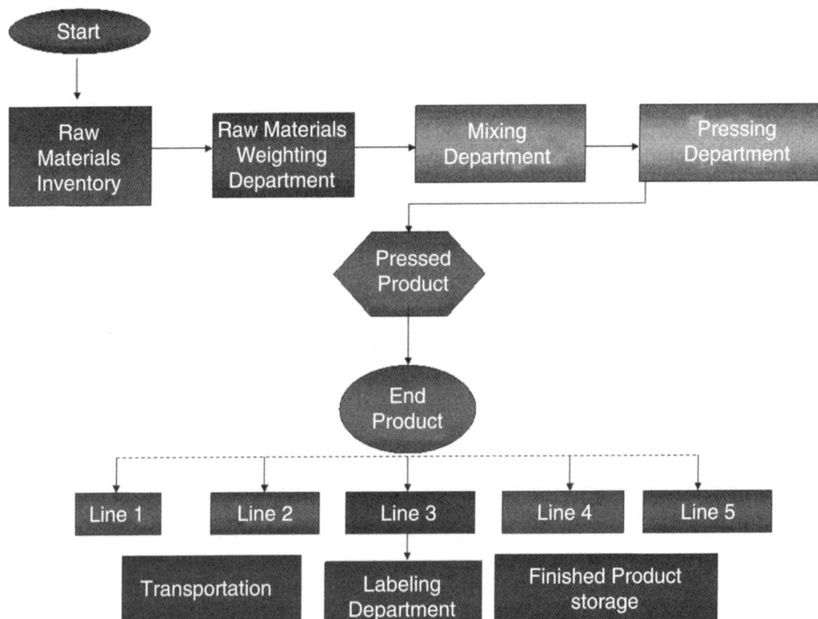

Figure 4.14   Product diagram

transform it into a special paste, to vulcanize it in the Bambury, and to shape it with a specific formula selected 180 possibilities the firm currently handle. The laboratory is used to test several formulas, and to make sure they complied with the American Society for Testing and Materials (ASTM) standards.

Despite the fact that the firms working in this market seem to be disconnected, they interact trough the input market that remains the same for all them. Precisely, it is through ICTs that participating firms negotiate raw material (melted rubber) with external providers or purchase remaining product/unused inputs from big companies. Competition among firms is not severe because the market is characterized by an extensive number of niches that allow every firm to catch its own sub-market.

5.2.3   ICTs and the Hulera Costarricense

ICTs had played an important role in the progress showed by this firm over time, being an integral part of management, distribution, contact suppliers and potential clients, although they are not yet fully incorporated in the production realm. Another high point of this firm is the modern laboratory for simulation and ASTM testing purposes, being the only one of its kind that belongs to a national firm.

The first fixed phone was introduced in the firm in 1950, being the only ICT used in the company for 36 years. In 1986 the firm bought its first computer for accounting and administrative purposes. By that moment, just a few number of companies in Costa Rica had a computer incorporated at any level. In 1987, a digital network was introduced to manage inventories as a primary technological network. By 1990, a Novell network interconnected several working stations and coordinated most of the firm's activities. In 1991, a DOS intercom accounting module was introduced in the firm.

After almost 10 years without a significant technological shift, in the year 2000 the firm connected to Internet, allowing access to new opportunities for the firm's best interest, such as new information about providers, access to raw materials markets, product sales, small necessary quantities of specific goods and inputs and access to new, improved and modern machinery. Actually, the firm finds itself facing a dilemma regarding the ICTs situation. Although it has access to Internet, they still keep the Novell network, being time of changing to a modern NT network.

Part of the changes experienced in the technological field is due to the international collaboration that the company shares with other firms in areas such as design, training and purchasing of raw materials. Transfer of technology was the highest-ranked benefit coming from international partnership, followed by brand name. Benefits are mostly related to the use of ICTs and can be perceived in the contact of clients and international providers. Finally, results of the survey show that the current competitive situation of the firm is below the rest of the sample. Increased competition and reduction in demand for rubber products have eroded the position in the market. The firm lacks institutional support mainly because the sector does not have a chamber.

5.3 Aceros Vargas

For some SMEs, competing in international markets becomes a challenge. Now, for SMEs to be competitive in international markets, experts such as Porter (1990) studied the need to constantly innovate through improved technology in both production and management dimensions. According to Pelupessy, Gereffi and Kaplinsky (cited by Gomez, 2002) SMEs have the ability to interact in international markets if they aim at small and specific market niches. Aceros Vargas, our third case study, is an example on how a SME can break those fears and can achieve substantial benefits from ICTs and a well-planned strategy.

5.3.1 Historical background

Aceros Vargas is a firm of the metallurgy industry that manufactures steel pieces for its clients. During the 70s, the Costa Rican industrial policy focused in developing instruments to support industries and to stimulate new businesses. It was this kind of policies that led to the foundation of Aceros Vargas. The company emerged when the government nationalized the taxi services and acquired the family taxi company, giving the owners a steel company as a compensation for the previous loss.

Aceros Vargas was founded in 1974 by Mariano Vargas Malavasi and started working with 3 employees. In that year, Roberto Vargas, son of the owner, traveled to Brazil and got a Master's degree in the analysis of steel. Time later, Roberto Vargas would take over the firm in 1985 after his father retired and with a renewed perspective of the steel industry, incorporated several changes that modernized the firm. In 1992, with the financing support the Banco de San José, the company began the construction of a new and improved building. The new edification was built considering the needs of a corporative, professional and technical approach of the operation, aiming towards time advantages in the delivery schedule.

By 1994 a revolutionary change occurred in the firm. The steel cutting business was, until that year, a labour-based activity in which the steel was cut by hand in a very imprecise way, creating a large amount of waste. In order to reduce that problem, Mr. Vargas created special software that transformed the company into the first firm in Latin America with a fully automatic cutting, storage and delivery system controlled by computer software. The implications of this programme have been very important to the firm and will be explain later.

5.3.2 Aceros Vargas context

The context of SMEs in Costa Rica turns around finding new ways to approach the markets with the technologies available provided by public institutions and training programmes. Aceros Vargas has been able to create a business related to high-tech strategies, reaching out the customers.

Although the activities developed by this company are classified as heavy industry, the general manager Roberto Vargas considers that the product they provided can be categorized in two different branches, the steel cutting business that can be consider as a product and the services related to this product, such as certification procedures. Aceros Vargas can be considered as a heavy industry that generates medium value-added products and high value-added related services. On average, imports represented 35 per cent of total sales during 2004, but this figure could be undervalued if one considers that total sales include services, not only steel products. Total sales turnover and imports for 2003 and 2004 are presented in Figure 4.15.

What makes this firm highly remarkable is its ability to take advantage of ICTs to create a unique automatic process, breaking the existing rule in Costa Rica where almost all heavy industries are labour intensive activities, not capital intensive ones. Even more interesting, the firm only has 15 employees, being the smallest of all the cases considered in this chapter.

Providers are mostly European and they are vital to the survival of company by proving high quality steel for its operations. Historically, raw materials mainly come from Sweden, representing 80 per cent of total purchases, while Germany and Austria came in the scene with the remaining 20 per cent. Most of the imports made by this firm are directly related to the cutting process.

Figure 4.15 Total sales and imports for Aceros Vargas

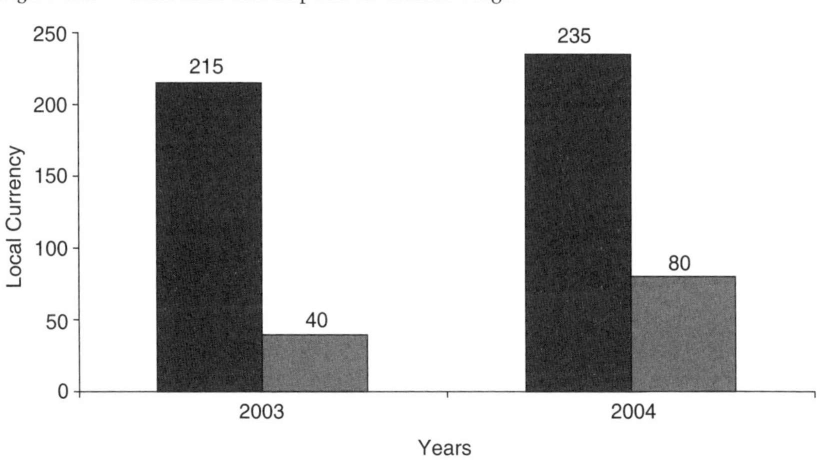

Source: Aceros Vargas.

Right now, the industry export to Central American countries using both Internet and a fixed telephone line. However, one of the key barriers for expanding operations is the size of the market. During the last years, continuous growth over time made the company to reach a limit, not only at the domestic level but also at the Central America level. The strong position of the firm is eroded due to its failure to increase sales in a sustained way. For that reason, the factory has since 2 years ago, the General Manager outlined an interesting project for installing a 'virtual mall' where clients can purchase not only the steel pieces but some other goods that are members of the mall.

The overall idea is to design a virtual mall so every customer can access the webpage and analyse the information on steel pieces and complementary services. For this plan to be successful the technological gap presented in terms of connection and general Internet infrastructure must be closed. Current conditions do not allow for fast data transmission, making impossible to sell goods using Internet. The information can be displayed, but the transaction cannot be executed. Although the idea is highly attractive and potentially profitable, the project cannot be executed due to that limited capacity. Preliminary estimations have shown that, with the project, Aceros Vargas can multiply by five its current levels of sales.

5.3.3 Software development and competitiveness

This firm has four working areas related to the steel selling business (Fig. 4.16). The first one deals with steel processing, mainly cutting. It is in this process that most of the ICT innovation is found because the cutting stage comprises the application of the special software developed in the firm. The software has been adapted to the particular conditions of the company and provides quick information about the availability of each of the pieces manufactured by the company, being an advantage for the interaction with clients. The computers share a common connection with all programmes integrated with normal Microsoft nets.

The second activity corresponds to the direct steel import and selling. The third and fourth activities are the computerized assessment of pieces and the industrial heat treatment, this last one incorporated ISO-9000 quality standards. Now the former activities are commonly related to the first one.

Computerized assessment of pieces is a service provided by the firm to advice the customer on the characteristics of original piece and the capacities able to support. The use of this service is provided both to customers that have already purchase the steel part in the firm or in some other place.

Finally, industrial heat treatment can also be performed by this firm on pieces acquired in this enterprise or for pieces purchased on other firms. This industrial treatment is used to overheat the piece so it can be manipulated and formed at convenience of the client.

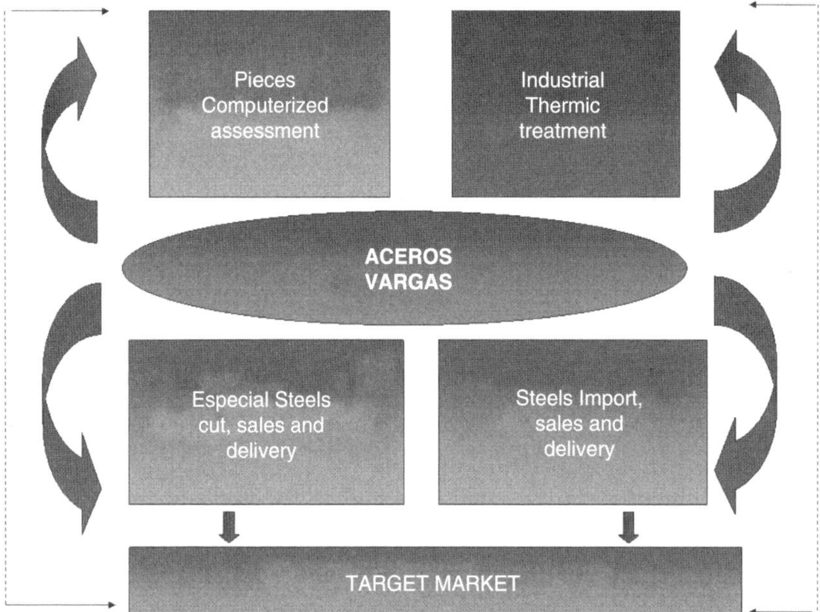

Figure 4.16   Aceros Vargas market structure

The tailor-made software created a competitive advantage for Aceros Vargas because it allowed the company to measure, record and store each one of the steel pieces generated during the production stage. Similar firms in the market, though, just cut the pieces but do not have an adequate control over the inventory, something that delay the possibility of giving prompt responses to the clients. The software dramatically reduces the time of response by keeping the record of all cuts performed and therefore a complete registration of what is stored in any moment of the day. The system, thus, facilitates sales and tracks any shortage of pieces, while provides information to the management about the credit situation of the purchaser and the total cost of the piece (transportation included). All these steps are performed in real time considering a really fast market. The software transformed the remaining pieces in 'assets', frequently updating the inventory. Aceros Vargas has reduced client's requests of pieces to instant attention and consequently has acquired a structural advantage over local and foreign firms dedicated to this activity.

Foreign competition located in the country has gone bankrupt due to this competitive advantage. Three firms have already experienced the intense competition present in this sector. Only one foreign firm, from Colombia, remains in the market since 2000. Spanish firm SEXCO left the market in the same year.

The competitive advantage created with the special software (Fig. 4.17) has been perfectly complemented by the high level of human resources the company has. Out of the 15 employees currently working with Aceros Vargas, three are engineers (20 per cent) that provide excellent support and new ideas to the overall system. The firm makes use of all ICTs except those related directly for production such as CAD, CAM, CAE, FMS and numerically controlled machinery, but this can be overcome by the improved inventory system and several innovated applications.

The reasons to use ICTs are various but mostly because of the design flexibility, unit cost reduction, production efficiency, opportunity to provide complementary services and efficient management control over the inventory. Direct consequences of using ICTs were summarized in productivity increases, deskilling, time advantages, management control and post-sales support. Constraints to access ICTs are mostly centred in the Internet connection and speed (as previously mentioned) and communication in general. Other obstacles to access ICTs are the lack of qualified human resources in the metallurgic sector and infrastructure vacuums.

In terms of competitiveness factors, the firm does not present a very different profile from the other surveyed SMEs. The difference relies in the fact that as part of the metallurgic sector (traditionally a capital-intensive sector), its competitive advantage depends on R&D, quality product, flexibility design, brand name and the delivery schedule thanks to its technological advances.

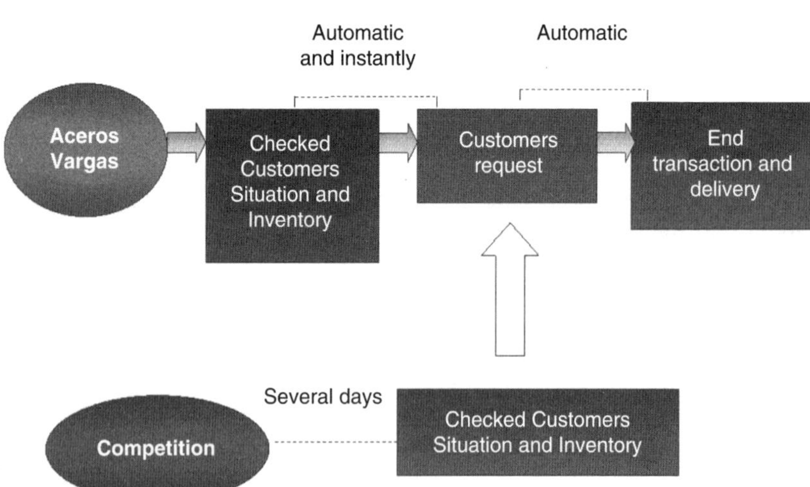

Figure 4.17   Software product diagram

6   Summary and conclusions

The study can be summarized in three parts. First we have evaluated economic performance of Costa Rican SMEs and its contributions to the national economy. Despite their substantial contributions to national economy, they have not been given due importance until recently. Second part of the study delineates the policy initiatives that have been taken since mid-1980s. Most of the initiatives were exclusively meant for SMEs. Third part of the study focuses on the innovativeness of SMEs in general and identification of factors that resulted in varying degree of ICT adoption. In order to identify those factors, a survey of 68 firms was conducted during July 2004 and February 2005.

Although SMEs contribute significantly to the national economy in terms of jobs and exports, public efforts to improve their competitiveness have been erratic or weak, so the progress towards that objective seems to be small. In terms of industrial policy, the experience of the last 20 years suggests that an overall industrial strategy should be developed with a long run perspective, more than the 4 years of the presidential term. Support programmes aimed at SMEs should be overviewed through administrations and become a long-term plan focused on cluster economies based upon technology improvements and R&D to develop new and improved ways to achieve productivity and competitiveness.

SMEs are relatively young companies managed by middle aged, highly educated managers. Common strategies to compete are quality and product differentiation in specific niches. Training is important but R&D has very few rooms in the SMEs agenda. ICTs are highly significant factor in the performance of SMEs. Although the country has been able to provide extensive coverage of basic technologies in the last 40 years (fix lines, for instance), there is some lagging behind in comparison to other countries with respect to modern technologies such as Internet (coverage and speed). Although the package of Internet services provided by National Institute of Electricity allows firms to use basic functions, lack of an adequate infrastructure is impeding SMEs to advance in the use of Internet for productive, marketing and sales purposes. In general, improved performance in productivity, time advantages, design flexibility, reorganization and management is associated with an extensive use of advanced ICTs, supporting the idea that ICTs are essential for enhancing the competitive profile of the firm.

It was not possible to investigate the role of policy initiatives in the context of the performance of SMEs using statistical techniques. However, firm level data allowed us to identify and analyse factors that discriminated advanced ICT using firms from the rest. The ICTs that were included in the analysis are email, Internet, portal, web enabled technologies, MIS, CAD/CAM, CAE, FMS, and CNC. Firms were categorized into three groups, namely: low level of ICT using firms, moderate users of ICT, and advanced ICT using firms. Classification was done by a statistical technique called cluster analysis. Clustering of firms based on their intensity of ICT use was imperative as the sample had to be divided into reasonable number of distinctive groups. Subsequently, discriminant analysis was used to identify factors that discriminated different levels of ICT using firms.

Factors representing, entrepreneurship, international orientation, causes and consequences of ICT use, sources of competitiveness, and knowledge acquisition opportunities were used in the analysis. The results suggest that GMs knowledge base and academic background emerged as an important discriminant. The study finds that skill intensity of advanced ICT using firm was higher than the rest. International orientation of firms also discriminated three groups of firms. Finding also show that GMs that attributed higher importance to ability of ICTs in providing better management control and useful market information adopted more advanced ICTs. The study finds evidence to support the argument that the adoption of ICTs results in productivity gains and efficiency in production processes. We have not been able to evaluate performance of SMEs as a result of the adoption of ICTs due to the lack of data.

We conclude that the government needs to encourage and provide institutional support to SMEs to participate in international markets so that they remain globally competitive. Lack of this type of support could be attributed to the present state of low level of globalization of Costa Rican SMEs. Greater participation in global market might enable firms to increase their contribution to the national economy.

Appendix Table 4.1   Economic performance (in millions colon)

Sector distribution	Number of workers	Total sales per year	Imports per year	Exports per year
Manufacture of wood products and wood and cork	50	150	105	60
Manufacture of basic metals	30	360	Na	Na
	13	225	Na	Na
Manufacture of paper and paper products	17	36	10–15	Na
Recreation, cultural and sporting activities	10	25	Na	Na
	40	420	189	126
	50	550	500	200
Manufacture of textiles and apparel	70	180	Na	Na
	60	168	210	168
Manufacture of chemicals and chemical products	30	150	Na	Na
	70	2940	840	1890
Manufacture of rubber and plastic products	18	120	30	2,4
	24	1000	Na	700
Manufacture of food products and beverages	10	32	Na	4,2
	42	1500	Na	675
	22	480	315	Na
	40	4800	150	16,8
	26	200	15	12
	20	600	Na	390,0
	40	261	104	260

Appendix Table 4.2   Learning modes

ICTs	Average score		
	Learning by doing	Training by other institutions	Formal training
MIS	4.20	3.85	3.50
Portal	4.24	3.79	3.61
Web enabled	4.15	3.94	3.59
Internet	4.17	3.75	3.65
Email	4.18	3.73	3.60
CAD/CAM	4.00	3.33	4.00
CAE	4.00	3.33	4.00
FMS	4.04	3.81	3.58
CNC	3.97	3.63	3.48
Total	4.14	3.69	3.59

Appendix Table 4.3   Variables included in the analysis

Variables	Description
1. Entrepreneurship	
MD_EDU	Academic background of general managers
ENG	Workers holding degree in bachelor of engineering
2. International Orientation	
FORN_COLL	Foreign collaboration (binary variable)
3. Causes of ICTs use	
MGMNT_CTRL	Opinion of GMs on contributions of ICTs in better management control (five-point scale)
MAR_INFO	Provision of market information due to use of ICTs (five-point scale)
EFFI_PROD	Contributions of ICTs in production efficiency (five-point scale)
4. Consequences	
LEAD_TIME	Adoption of ICTs results in lead time advantages (five-point scale)
PROD_GAIN	Use of ICTs contribute in productivity gains (five-point scale)
FLEX_DESIGN	ICTs enable flexibility in product designs (five-point scale)
5. Competitiveness	
PROD_QUAL	Role of product quality in competitiveness (five-point scale)
DEL_SCH	Importance of delivery schedule in competitiveness (five-point scale)
6. Human resource development	
LEARN	Role of learning by doing in knowledge acquisition (five-point scale)

Appendix Table 4.4   Acronyms

ACORDE	Costa Rican Association for Development Organizations
BCR	Costa Rican Bank
BNCR	National Bank of Costa Rica
BPDC	Popular Bank of Costa Rica
CAATEC	High-Tech Advisement Commission for Costa Rica
CACIA	Food Alimentary Industry Chamber
CADEXCO	Chamber of Costa Rican Exporters
CAPROSOFT	Chamber of Software Producers
CEGESTI	Center for Technological Management and Industrial Information
CICR	Chamber Costa Rican Industries
CRECEX	Chamber of Foreign Industries
FUCODES	Costa Rican Association for Development
FUNCENAT	Foundation for High Technology
FUNDES	Foundation for sustainable development among SMEs
INA	National Training Institute
INCAE	Central-American Institute of Business Management
INTECO	Costa Rican Institute for Technical Standards
ITCR	Technological Institute of Costa Rica
MICIT	Ministry of Science and Technology
PROCOMER	Commerce Promotion Institute
UCR	University of Costa Rica
UNA	National University of Costa Rica

5
New Technologies and SMEs in Jamaica[12]

1　Introduction

Small and Medium-sized Enterprises (SMEs) are the growth engines of developing economies as they generate employment opportunities, and contribute to overall gross domestic product. However, there are a number of constraints limiting the growth of SMEs such as inadequate credit facilities, infrastructural and technological problems, which hamper the development of this vital sector of the economy and limit the growth, global competitiveness and the formation of linkages in the global commodity chain.

Governments in several developing countries have taken initiatives to retain/strengthen competitiveness of SMEs in the era of globalization, which is characterized by global business environment. These measures could be technological up-gradation support, new opportunities for human resource development, and marketing support. Technological support is not limited to making SMEs aware of advanced technological development, which has taken place in the world market but also to help them in acquisition of new technologies. One of the possible ways to encourage firms to adoption of new technology is through the reduction of cost of accessing such technologies. This can be achieved by providing soft loans for importing new technologies. For export oriented SMEs a reduction in custom duties might encourage adoption of new technologies.

Support to SMEs in the era of globalization has become imperative due to entry of Multi-National Corporations (MNCs) in developing countries. It is well known that MNCs are equipped with several tangible and intangible assets. These assets could vary from efficient production technologies to brand name advantages. In order to compete even in the domestic market

---

12　Abridged version of the chapter has been accepted for publications in ICFAI Journal of Knowledge Management.

SMEs need to upgrade their business processes, which is virtually imposs-ible for SMEs without active support of the government. This is primarily because neither SMEs have enough resources at their disposal to invest in new technologies, nor do they have enough risk absorbing capacity.

Governments in developing countries are aware of these problems and many initiatives have been taken to encounter the onslaught of globaliza-tion. However, one-way flows of policy initiatives are not sufficient for their success. There has to be willingness and eagerness to adopt new tech-nologies in SMEs. The absence of such motivations might lead to misuse of policies and government resources. There have been several studies (Lal, 2004; Lee, 2003; McFarlane, 1997) that examined the policy measures initi-ated in India, Malaysia, and Jamaica and their implications on the perfor-mance of SMEs. However, most of the earlier studies particularly in Jamaica have been based on macro-data. This study on the other hand makes an attempt to examine performance of SMEs in Jamaica by involving all the stakeholders that are responsible for the growth of SMEs. Information on the policy initiatives were collected from concerned institutions/ministries and subsequently firm level data were collected to investigate the impact of such policy measures and initiatives taken by entrepreneurs on the perfor-mance of firms.

In Jamaica, fiscal and credit support to SMEs are available in terms of lending institutions – some of which offer additional training and techno-logy support. However, the extent of usage of these credit lines and techno-logical support are not known since most SMEs remain outside the formal sector of the economy. Earlier studies have found evidence of the structural change in ownership patterns, sales volume and the business environment of SMEs in Jamaica. In this study, we evaluate the competitiveness of SMEs in Jamaica and their ability to use Information and Communication Technologies (ICTs) to enhance global competitiveness and add value in the global commodity chain.

It may not be possible to cover all the aspects of SMEs in a single study. However, an attempt has been made to identify and analyse impediments and driving forces in the adoption of new technologies, which are by and large led by ICTs. The main objectives of the study are: to evaluate the degree of ICT adoption in SMEs in Jamaica, an examination of policy initiatives aimed at encouraging the adoption of new technologies, and to analyse impact of the adoption of new technologies on the performance of SMEs. Although leap-frogging of the adoption of ICTs is very much possible, the successful use might be affected by the availability of other prerequisites such as persons with appropriate skills, legalization of e-business, and proper local, regional and national technological infrastructure. Hence it is considered vital to present historical fact about evolution of SMEs in Jamaica.

Remainder of the chapter is organized as follows. It was thought of crucial to understand the historical growth of SMEs in Jamaica. Section 2 is devoted for that. Institutional support provided to SMEs is discussed in Section 3 while evolution of structural and organizational changes is presented in Section 4. The methodology, theoretical framework, and hypotheses are discussed in Section 5. Statistical results are presented and analysed in Section 6 whereas conclusions and main findings are summarized in Section 7.

2   Historical growth of SMEs in Jamaica

The importance of SMEs and their contribution to economic development during an era of globalization have been continuously stressed by various governments and private sector commentators. In the Caribbean region and most of the developing world, experience has shown that SMEs are the vehicles of entrepreneurial creativity through important contributions to total manufacturing output, the implementation of new technologies and the provision of much needed jobs and the creation of employment opportunities. In addition, SMEs cater to the basic needs of the populace and are less dependent on imported materials than larger companies, and constitute more flexible sources of innovation (McFarlane, 1997).

Although, there are no universal definitions of micro and small-scale enterprises, most of the privately-owned firms in Jamaica fall within the realm of micro-enterprises. In Jamaica, micro and small enterprises (MSEs) predominate over small and medium-sized enterprises (SMEs) and are defined as businesses employing less than ten persons with an annual turnover of less than USD 25,000. On the other hand, SMEs are defined as businesses employing less than 50 persons with turnover between USD 125,000 and 1 million. No matter how defined, the growth of micro and small business in Jamaica has been phenomenal since 1980.

The 1960 Survey of Business Establishments was the first survey of small businesses in Jamaica but it was rather limited in scope to only the manufacturing and construction sector leaving out the distribution and service sectors. On the other hand, the 1983 Small Establishments Survey covered all non-agricultural micro and small businesses.

There were similar surveys in 1990 and 1992. The 1997 report by McFarlane (1997) on the growth of MSEs in Jamaica indicated that the growth of small businesses was faster in the 1980s than during the 1990s. A growth rate of 141.6 per cent was recorded between 1983 and 1990 or annual growth rate of 13.43 per cent whereas during 1990–1996, the growth rate was 4.96 per cent or an annual growth rate of 0.81 per cent. Over the period, there has been a shift in numbers of firms from the industrial sector to wholesale and retail trade distribution. The estimated sales volume also increased to about Jamaican $ 48.594 billion of which wholesale and retail comprised 43 per cent (see Table 5.1).

Table 5.1 Comparison of total sales in micro and small business with total gross output in 1996

Industry	Annual sales ($'mil)	Gross output ($'mil)	Sales as % of gross output
Manufacture	6,090.0	107,474.7	5.7
Construction	404.5	90,405.3	0.5
Wholesale and Retail Trade	29,239.9	67,506.3	43.3
Restaurants and Hotels	4,647.8	19,996.3	23.2
Transportation, Communication, Storage	621.4	45,746.0	1.4
Finance and Business Services	4,931.8	34,037.5	14.5
Community and Social Services	774.8	5,971.1	13.0
Personal Services	1,880.9	2,675.3	70.0
TOTAL	48,594.1	373,812.5	13.0

Source: McFarlane (1997).
Note: Unit of measurement is Jamaican Dollar.

Table 5.2 presents distribution of firms by type of business and industry. It can be seen from Table 5.2 that own account-workers accounted for a significant proportion of all micro and small-scale enterprises in Jamaica and their percentage has increased from 47.2 per cent in 1992 to 72.9 per cent in 1996. Although the average firm size has increased from 1.38 to 1.76 during same period, the percentage of paid workers has decreased. We can conclude from Table 5.2 that there is greater tendency to employ family members rather than hiring people from labour market.

Table 5.3 presents the pattern of changes in employment in MSEs during the early 1990s. Although total employment in absolute terms has increased from 1992 to 1996 in MSEs, its share in total labour force has decreased from 18.3 per cent to 18.1 per cent during the same period. We can infer that rigidity in labour market has increased during this period. Data presented in Table 5.3 also suggest that percentage of own account or unpaid workers has drastically changed from 1992 to 1996.

Table 5.4 presents distribution of paid and unpaid workers in micro-enterprises by type of industry. It can be seen from the table that wholesale and retail trading firms constitutes a large portion of micro-enterprises. It is improper to comment on the change of industrial structure during the 1990s due to lack of data. However, the disappearance of export processing zone and other industrial clusters suggests that manufacturing firms might have declined in the last decade.

It is difficult to put a precise figure on the income generating activities of SMEs in Jamaica because most of them operate outside the formal sectors of the economy. In a study conducted between 1979–81 by the Michigan State University and University of West Indies, it was shown that over

Table 5.2 Distribution of micro-enterprises by business type and industry

Industry	Type of business						
	Own account workers	1–2 paid workers	3–4 paid workers	5–9 paid workers	Total	Avg. no. of workers	Sample size
1992							
Manufacturing	41.7	33.8	16.5	8.1	100.0	1.53	(480)
Construction	51.9	22.2	3.7	22.2	100.0	1.74	(27)
Wholesale and Retail Trade	60.9	28.4	6.4	4.3	100.0	2.16	(719)
Restaurants and Hotels	37.8	40.4	15.0	6.7	100.0	1.56	(193)
Transp./Conmmuni/Storage	45.8	36.1	10.8	7.2	100.0	1.34	(83)
Finance/Business Services	10.7	42.7	30.7	16.0	100.0	2.69	(75)
Community/Social Services	22.2	44.4	23.3	10.0	100.0	2.13	(90)
Personal Services	56.1	35.7	6.7	1.5	100.0	0.80	(462)
Motor car/Other repairs	30.0	35.8	24.3	9.9	100.0	1.80	(263)
TOTAL	47.2	33.9	12.7	6.2	100.0	1.38	(2,392)
1996							
Manufacturing	54.6	26.5	11.9	7.0	100.0	2.41	(326)
Construction	47.2	10.4	20.7	21.7	100.0	3.73	(22)
Wholesale and Retail Trade	82.2	13.2	3.5	1.1	100.0	1.42	(530)
Restaurants and Hotels	46.6	37.0	9.9	6.5	100.0	2.03	(214)
Transp./Conmmuni/Storage	69.2	16.4	6.1	8.3	100.0	1.49	(65)
Finance/Business Services	31.1	26.3	21.3	21.3	100.0	3.17	(46)
Community/Social Services	34.5	43.3	14.3	7.9	100.0	2.04	(73)
Personal Services	74.3	20.8	4.1	0.8	100.0	1.35	(599)
Motor car/Other repairs	54.5	33.3	8.9	3.3	100.0	1.99	(169)
TOTAL	72.9	18.8	5.6	2.7	100.0	1.76	(2,044)

Source: McFarlane (1997).

Table 5.3   Contribution of employment by MSEs

Labour force characteristic	1990			1996		
	Total labour force	Total in MSEs	MSEs as % of total labour force	Total labour force	Total in MSEs	MSEs as % of total labour force
Employed labour force	896,300	163,680	18.3%	959,780	174,010	18.1%
Paid employed labour force	509,275	113,290	22.2%	569,900	65,600	11.5%
Own account/ unpaid workers*	387,025	50,390	13.01%	389,880	108,410	27.8%

Note: *→ computed from the report.
Source: McFarlane (1997).

Table 5.4   Distribution of workers in micro-enterprises by type of worker by industry in 1996

Industry	Working entrepreneurs		Unpaid family workers		Paid employees	
	No.	%	No.	%	No.	%
Manufacturing	8,380	9.2	850	4.7	8,740	13.3
Construction	380	0.4	–	–	830	1.3
Wholesale and Retail Trade	56,760	62.6	14,680	80.7	23,770	36.2
Restaurants and Hotels	8,640	9.5	1,430	7.9	14,430	22.0
Transport/Communications/ Storage	1,200	1.3	20	0.1	1,090	1.7
Finance and Business Services	1,730	1.9	110	0.6	4,540	6.9
Community and Social Services	2,120	2.3	230	1.3	3,800	5.8
Personal Services	9,390	10.4	630	3.5	5,740	8.7
Motor car/ other repairs	2,050	2.3	230	1.3	2,660	4.1
TOTAL	90,650	100.0	18,180	100.0	65,600	100.0

Source: McFarlane (1997).

40,000 SMEs (comprising 35 per cent in the manufacturing and 50 per cent in the distributive trade) employed about 80,000 persons. Unfortunately data on growth of SMEs are lacking as the Jamaican Registrar of Companies does not have a database of SMEs. However, the value of gross output for the year 2003 was $ 145.3 million or 27.5 per cent of total sales from general consumption tax returns representing an increase of $ 5 million over the preceding year (PIOJ, 2003).

2.1   SMEs' needs in the Caribbean

The major challenges facing SMEs in the Caribbean region has been well articulated in several reports (McFarlane, 1997; Priestee, *et al.*, 1998). According to these reports, the major problems of SMEs in Jamaica lie in the lack of access to inputs and markets and the enabling environment for growth, ageing equipments and outmoded technology, shortage of skilled manpower, lack of raw materials and inadequate demand for finished goods. The main start-up problems affecting SMEs in Jamaica were identified as capital and credit, demand and supply problems, building and machinery problems (Table 5.5).

Some of these problems are related to the size of the industry and the sector of operation. For instance, low market demand was predominant with the restaurant and hotel industries whereas shortage of spare parts affected the transportation industry more than others. High taxation and inflation were of a problem to wholesale and retail services. According to the Inter-American Development Bank (IDB) report (Priestee *et al.*, 1998),

Table 5.5   Main start-up problems experienced by micro-enterprises by industry – 1992

Industry	No problem	Capital & credit	Demand problems	Supply problems	Building	Other incl. machinery
Manufacture	34.4	41.7	15.0	13.3	5.4	9.8
Construction	51.9	18.5	18.5	–	–	3.7
Wholesale/Retail Trade	42.8	34.1	12.0	7.6	4.5	1.8
Restaurants and Hotels	48.2	25.9	13.0	2.1	6.2	1.0
Transport/Comm/Stor.	51.8	28.9	6.0	2.4	–	3.6
Finance/Business Services	52.0	22.7	17.3	2.7	6.7	–
Community/Social Services	43.3	27.8	12.2	6.7	10.0	5.6
Personal Services	41.6	28.8	21.6	4.5	6.1	5.2
Repairs	28.5	37.6	21.7	6.1	9.5	6.5
TOTAL	38.8	33.4	15.6	6.8	5.7	4.7

Source: McFarlane (1997).

the first major challenge facing SMEs in Jamaica resulted from the polarized nature of industry in the region. There are a small number of large firms and a very large number of informal, family-owned small and micro-enterprises. According to the IDB 'this structure creates a potential growth area for SMEs'. On the other hand it also imposes some rigidity to the productive structure by impeding access to factors of production and offering limited opportunities for networking and the creation of industrial clusters. Other identified constraints included the large fixed costs, absence of economies of scale and the high cost and lack of access to financial and non-financial services. The SMEs have limited access to long-term financing and other financial services and there are the high transactional and due-diligence costs. The report also identified the impediment to growth arising out of high tariffs and high legal costs, which do not proportionately affect large enterprises. In terms of non-financial barriers – there are difficulties with training, labour, technology, consulting services and other inputs that inhibit SME competitiveness.

There are also demand and production constraints as a result of the environment in which SMEs operate in Jamaica. In the order of priority, the most serious of these problems are capital and credit problems (44.6 per cent), low customer demand (34.1 per cent) and customers collecting goods without payment (30 per cent), those related to raw materials and/or supplies (11.1 per cent) and failure of customers to collect their goods (10.9 per cent) (Priestee *et al.*, 1998). Other problems accounting for less than 10 per cent of the constraints were theft and vandalism, shortage of skilled labour and the quality of raw materials.

3 Institutional support for SMEs in Jamaica

Jamaica's National Industrial Policy (NIP, 1996) sought to broaden the market for MSEs through export promotion initiatives and calls for efforts to link small businesses with larger firms through subcontracting relationships. Private sector support for small firms is available in the form of supporting institutions such as the Jamaica Chamber of Commerce (established in 1885), the Small Business Association of Jamaica (established in 1974) a private sector organization of Jamaica, and Jamaica Export Promotion Organization (established in 1983). These organizations provide financial and non-financial services to members, maintain a register of SMEs, provide training and afford opportunity for SMEs to network and interact with other members. However, there are probably many more SMEs outside of the network, paying no taxes and of unknown income.

3.1 Government of Jamaica support for SMEs

In Jamaica, public sector support for MSEs as well as SMEs are channelled though the Ministry of Science, Commerce and Industry and its agencies and these have been on the increase although it is not clear whether these inputs have been linearly related to outputs. Total disbursements by all agencies was J$ 936 million in 2003 (PIOJ, 2003). Financial assistance to micro and small businesses occur through five main bodies – the National Development Foundation of Jamaica which provides regular loans of $ 20,000 to $ 3.5 million, the Opportunity-Knocks (OK) programme which provides loans of about $ 40,000 to individuals, the self start fund, and the Micro-Development Investment Agency. In addition, the Jamaica Exporters Association and the Caribbean Regional Programme for Economic Competitiveness provide non-financial support occur through computer training, management in financial management.

The Jamaica Business Development Centre established in 2001 has two main objectives – to developing a network of Business Service Providers (BSPs) and strengthening the BSPs as viable commercial entities. In addition, multilateral institutions such as IDB, United Nations Development Programme (UNDP), and United States Agency for International Development (USAID) have all provided support for SMEs in credit financing and capacity building since the 1990s. IDB support for SMEs has been through three related institutions. These are:

The Inter-American Development Bank – an international financial institution which supports accelerated economic and social development in Latin America and the Caribbean through the use of capital funds to supplement private investment and to provide technical assistance for the preparation, financing and implementation of development plans.

The Inter-American Investment Corporation (IIC) – IIC began its operation in 1989. Its operations are directed at providing equity investments through loans for projects for which traditional financing would not otherwise be available under similar terms. As of 1997, the IIC has participated in 180 transactions totaling 450 million through eight venture capital funds for a total of 35 million. The IIC also adds value to its loans by offering fee-based financial advisory services to help SME structure financially-sound projects.

The Multilateral Investment Fund – it undertakes activities which promote broader private sector investment in the economy by providing funds for equity and quasi-equity funds for intermediary institutions which support small-scale enterprises through its three windows. Window 1 – public policy and regulatory framework, Window 2 – Labour force training and Window 3 – finances projects which promote private enterprise – including provision of financial and non-financial services.

Although the support has been mainly in the area of financial services through direct and indirect financing (lending through intermediaries and investment funds), there are also some non-financial services provided to SMEs. By 1998, IDB had financed 16 multi-sector credit services through financial institutions worth $ 2.9 billion and 2 global credits for a total of $ 52 million. Non-financial services are provided through support for sectoral regulatory reform, strategy, enterprise development services and technology support.

The IDB support to SMEs also provides specific subprogrammes for scientific and technological needs of SMEs, Specifically they provide financing for technology-related programmes as reimbursable and no-reimbursable loans.

3.2   The United Nations Development Programme

The UNDP support for SMEs is mainly linked to its overall goal of eradicating poverty. Direct and indirect support to SMEs is aimed at supporting those enterprises to launch into the global marketplace. The support is through three main components of institutional strengthening via the building of technical support for SMEs, human resource development (through training in business development, solar technology development, Computer-Aided-design/Computer-Aided-Manufacturing (CAD/CAM) technology development) and through networking by developing complementary productive networks through programmes such as the Jamaica sustainable development networking programme.

3.3   The United States Agency for International Development

The USAID through its USAID/Jamaica Bilateral programme 'The New Economy programme' 2000–2005 seeks to promote economic growth and protect Jamaica's unique environment through improving the business environment, capital financing reduction of key business constraints, improving business skills and competitiveness. It provides support to SMEs through the New Partnership Initiatives, which provide lease financing, venture capital and increasing private financing to micro and small enterprise.

3.4   Micro-enterprise Financing Limited (MEFL)

This is a programme undertaken with the financial support from the Government of Canada through Canadian International Development Agency (CIDA). It is also supported by the Bank of Nova Scotia and the Kingston Restoration Company to provide loans for micro and small businesses. The total volume of wholesale financing to MSEs in 2003 was $ 936 million compared to $ 799 million in 2002. The main vehicles for administering loans were through the National Export and Import Bank (EXIM) and the Pan Caribbean Financial Services Ltd. In addition, other wholesale institutions include the Development Options Limited (DOL) which disbursed $ 70.9 million in 2003 and Microfinance Development Agency (MIDA) which disbursed $ 215.8 million to SMEs in 2003. Other institutions for financial support for SMEs in Jamaica are the Pan Caribbean Financial Services Limited and the Development Bank of Jamaica.

4 Evolution of structural and organizational changes in SMEs

The ability of SMEs to compete globally depends on their access to resources such as finance, technology and management skills. This requires a shift in production from simple corporate strategies to more complex integration between SMEs and transnational corporations (UNCTAD, 2000). However, the strength of the linkages depends on the technological capabilities of local firms and a coherent policy framework to foster investment, competition, technology transfer and SME development. In the current global climate, many transnational corporations (TNCs) are shifting to more complex network of interaction with SMEs through backward linkages with suppliers, linkages with technology partners, forward linkages with customers and other spillover effects. All these linkages depend on the technological and partnership readiness of local SMEs and the ability of host country TNCs not to set up barriers by becoming true global players. At the same time, developing countries are challenged to provide enabling environments to promote linkages between TNCs and local SMEs in the framework of global commodity chain and production networks.

It is argued in an UNCTAD (2000) report that the three factors; namely: the existence of supporting public policies and measures which facilitate technology transfer and competitiveness, TNC corporate strategies which are conducive to local SME development, and existence of SMEs which are able to meet high TNC standards promote TNC-SME linkages. In addition, one of the critical areas which is vital to the promotion of global TNC-SME linkages is in the area of technology. There is evidence that SMEs in developing and transitional economies often lack an awareness of new technologies because of shortage of funds, skills and training barriers. Technology linkages for information sharing, problem sharing may lead local SMEs to gain access to global markets and improve their competitiveness, apart from productivity gains, factor – cost advantages, flexibility and other spillover effects-demonstration effects, human capital spillovers etc (UNCTAD, 2000). If SMEs are to succeed globally, they need access to information, training, technology, consulting services and other inputs designed to enhance their competitiveness. TNCs also require some stability in the micro and macro-economic climate.

The structural changes of MSEs in Jamaica during 1992–96 have been well documented in McFarlane (1997). In Jamaica, the spatial distribution of MSEs indicates that majority of MSEs are located in the Greater

Table 5.6   Geographic distribution of non-agricultural micro-enterprises

Geographic location	Number			Percent		
	1983	1990	1996	1983	1990	1996
Kingston Metropolitan Area	11,030	49,860	20,470	30.0	56.1	22.0
Other Urban Area	25,700		32,450	70.0		34.9
Rural Area		38,990	40,190		43.9	43.1
TOTAL	36,780	88,850	93,110	100.0	100.0	100.0

Source: McFarlane (1997).

Kingston, St Andrew and St Catherine areas where financial services and infrastructure for business development is relatively well developed (Table 5.6). However in 1996, 43.1 per cent of MSEs were located in rural areas while only 22.0 per cent were located in Kingston Metropolitan area. It seems a large number of firms moved from Kingston Metropolitan Area to other urban areas in 1996.

There also appears to have been a major urban-rural drift and sole proprietorship seemed to be the dominant mode of ownership with business partners drawn from family or unpaid family workers. Most of the structural changes involving the use of technology have grown apace with the available lending climate and the provision of non-financial services by various multilateral agencies. Although, these structural changes have not been specifically directed at enhancing global competitiveness of small firms, they may have facilitated linkages of global commodity chains. According to McFarlane (1997) there has been a consistent shift in the industrial components of the SME sector. In 1983, trade comprised 47.2 per cent of total businesses in the sector rising to 63.3 per cent in 1996 (Table 5.7).

While enhanced use of technology appears to occur in many small firms, the overall structure of SMEs has remained relatively stable over the years. Of interest to fostering collaboration and partnership with TNCs is the issue of stability of SMEs. Most SMEs in Jamaica operate within permanent

Table 5.7   Industrial classification of non-agricultural micro-enterprises

Geographic location	Number			Percent		
	1983	1990	1996	1983	1990	1996
Trade	17,370	49,070	58,910	47.2	55.2	63.3
Non-Trade	19,410	38,780	34,200	52.8	44.8	36.7
TOTAL	36,780	88,850	93,110	100.0	100.0	100.0

Source: McFarlane (1997).

residences, owned by the entrepreneur and there have been a significant increase in the number of own-account workers (especially in the distributive and wholesale sectors). In terms of longevity, some 27.4 per cent of businesses in 1992 had been in business for 4 years of operation whereas the corresponding figures were 48.7 per cent in 1996. On the other hand businesses in operation for 5–12 years were higher in 1992 (42.7 per cent) than in 1996 (33.7 per cent). The opposite was found for businesses operating for more than 12 years i.e., 19.9 per cent in 1992 and 17.7 per cent in 1996 (McFarlane, 1997).

The spate of technology use and improvement seem to have increased apace, although in 1997, it was reported that the average age of machinery was 20–25 years in Jamaica (Priestee *et al.*, 1998). The IDB has provided support for the scientific and technological needs of SMEs through several operations. They provided financing for technology-related programmes and research and development.

4.1   E-readiness of Jamaica

Specific indicators of a community's e-readiness as provided by the Harvard University Guide are: Information infrastructure, Internet availability and affordability, network speed and quality, hardware and software service and support, state of schools, workforce access, E-commerce and e-government. In an assessment of all these indices, the conclusion is that Jamaica ranks from fair to good (Table 5.8). Several indicators of e-readiness such as tele-density, internet penetration, and personal computer density are presented in the table.

5   Data sources, analytical framework, and hypotheses

In order to assess the impact of the adoption of new technologies as a result of several policy initiatives on the performance of small firms, we conducted a survey of few firms. Methodology and data sources have been

Table 5.8   Jamaica's ICT base (2002)

Variables	Data
Population	2,621,043
Main lines per 1,000 inhabitants	170
Number of mobile phones per 1,000 inhabitants	535
Number of personal computers per 1,000 inhabitants	54
Number of Internet users per 1,000 inhabitants	228

Source: World Bank Development Indicator Database (online): http://devdata.worldbank.org/dataonline

discussed in the first sub-section while in sub-Section 5.2 we depict theoretical framework and formulate hypotheses.

5.1   Data sources

The study is based on a primary data collected though a semi-structure questionnaire (instrument). During the period August–November, 2004, we applied the instrument on a total number of 60 SMEs chosen from a list of 175 firms obtained by Small Business Association of Jamaica. Most of the sample firms were situated in the Kingston Metropolitan area. We collected information on various aspects of firms such as background and history of firms, international orientation, ICT profile of firms, consequences of the adoption of ICTs, and the impact of new technologies on the competitiveness of firms.

Most of the firms, which participated in this study had access to and utilized information technologies such as Management Information System (MIS), email, and the Internet. However, only a limited number was able to use advanced ICT tools like Computer-Aided-Design (CAD), Computer-Aided-Engineering (CAE), Computer-Aided-Manufacturing (CAM), web-enabled technologies, Flexible Manufacturing Systems (FMS), Computerized Numerically Controlled machine tools (CNC). Majority of the firms surveyed did not report any major constraint to the use of ICTs apart from the cost of telephone and Internet access. They did not also report any major implications for labour as a result of ICT use. The perception was that ICTs conferred modularity and flexibility in design and manufacturing of products.

5.2   Analytical framework and hypotheses

Analytical framework linking the adoption of new technologies and other factors is presented in Figure 5.1. This is a general framework depicting the nature of association between firm-specific strategies with regional and national policies. An econometric analysis of regional and national policies is not possible. Hence they are not included in the hypotheses. Based on the theoretical framework depicted in Figure 5.1, the hypotheses of the study are formulated in the following sub-sections.

5.2.1   Hypothesis I

As can be seen from the theoretical framework, conduct and performance of firm are subjected to several factors – most important being the entrepreneurial abilities of the owner, which in most of the cases is managing director (MD) in SMEs. It is the entrepreneur's knowledge base that drives innovations at the firm level. Innovation in SMEs invariably means the adoption of new technologies as inventing new markets and new products are usually beyond the capabilities of SMEs. Although academic background can be used as a proxy of knowledge base of MDs, we have used age

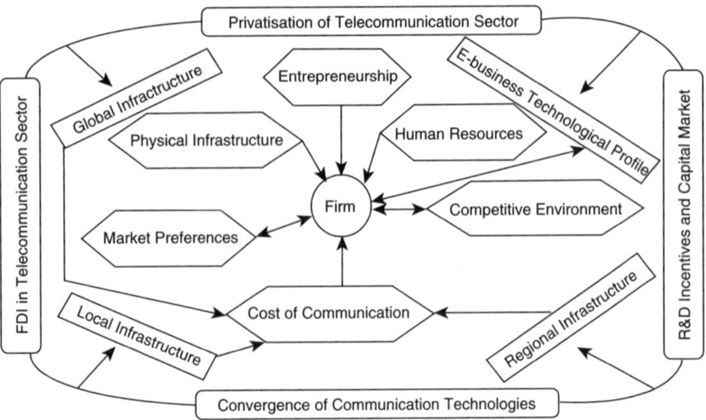

Figure 5.1   Analytical framework

of the MDs as a proxy of knowledge base. This is because earning of an engineering degree or advanced technical training several decades ago does not necessarily mean the person remains knowledgeable about the latest technological development especially about ICTs. On the other hand, a person who has recently earned advanced technical degrees is expected to be knowledgeable about application of ICTs as due emphasis on ICTs and their applications is given in any technical training programme in recent times. Hence, it was considered that age might be a better proxy of knowledge base.

Along with innovativeness, other factors that can result in better performance of firms could be skill intensity of workforce, size of operation, international orientation, physical and technological infrastructure. We have used number of workers that were using ICTs in firms as a representative of skill intensity. This is because one of the objectives of the study is to investigate the consequences of the intensity of the adoption of ICTs and intensity of adoption depends on the number and ability of its users. Total workforce has been used as a proxy of size of firms while collaboration with foreign partner has been used as a measure of international orientation. Cost and speed of communication and Internet speed have been used as proxies of technological infrastructure. We hypothesize that performance of firms, which is represented by sales turnover, is influenced by international orientation, size, quality of workforce, and availability of technological infrastructure at a competitive price.

5.2.2   Hypothesis II

Second hypothesis is related to the factors that influence the adoption of new technologies. As depicted in the theoretical framework, competitiveness, market preferences, availability of technological infrastructure, availability of human resources with appropriate skills, and market preference influence innovativeness of firms. It is not possible to test the association of intensity of ICT adoption and all the factors discussed above due to lack of data. However, we test and hypothesize a positive relationship between the intensity of ICTs adopted by sample firms and skill intensity, and availability of technological infrastructure.

Best way to test both the hypotheses would have been a simultaneous equation framework that would have captured causal relationship among many variables. However, that is not possible due to non-availability of data and sufficient number of degrees of freedom. Hence, we have used a neo-classical production function approach to test the first hypothesis and simple correlation analysis to test the second hypothesis.

Several production functions have been proposed and used in the literature. For instance:

EQ1. $Y_t = AK_t^{1-\alpha}(NX_t)^\alpha$ Solow (1956)

Where X_t denotes technical progress, which is assumed to expand at a constant rate

EQ2. $Y_t = BK_t^{1-\alpha}(NH_t)^\alpha$ Lucas (1988) discussed in Rebelo (2001)

Where H_t represents the level of human capital of the representative agents in the economy

EQ3. $Y(t) = K_1(t)^{\alpha 1} K_2(t)^{\alpha 2} \ldots K_m(t)^{\alpha m} [A(t)L(t)]^{(1-\sum_{i=1}^{m} \alpha i)}$

Mankiw, Romer and Weil (1992)

Where K_i denotes capital of type i (i=1,2,....m), L labour, A the state of technology.

The model shown in EQ3 is discussed in Pohjola (2001). Authors argue that in addition to investment in physical capital the basic Solow model is extended to include investment in human capital and in ICTs. The model which is used in this study is based on EQ3. We have used various costs associated for accessing ICTs as proxies of capital investments. The actual model specification is as follows:

$\text{Log}(Y_i) = \alpha + \beta_1 K_i + \beta_2 ICT\_SCR_i + \beta_3 H_i + \beta_4 O_i + \varepsilon_i$ i=1,2,3....................m firms

Where Y_i represents sales turnover, K_i represents various costs associated in using new technologies, ICT_SCR_i represents the state of technology, H_i represents human development resource variables, and O_i represents residual factors such as international orientation of firm i.

6 Methodology and statistical analysis

In this section we analyse data collected from sample firms. Data were analysed in multivariate analysis framework. Analytical framework used in this study is discussed in sub-Section 5.2 while profile of sample firms is presented in the following sub-section while statistical results are analysed in sub-Section 6.2.

6.1 Profile of sample firms

This section deals with the characteristics of sample firms such as product mix, age of firms, type of ownership, skill intensity of workers, and the initiatives taken towards upgradation of technology and consequences of the adoption of new technologies.

6.1.1 Products mix

Firms classified by the type of industry are presented in Figure 5.2. It can be seen from the figure that highest percentage of firms (28 per cent) belong to service sector. The type of services provided these firms include consultancy, financial, repair works, architects, printing, and insurance. The second largest percent (22 per cent) of firms were engaged in trading activities. These include auto and auto components trading, sale of electronic equipment such as cell phone and computer hardware, and stationery suppliers. Eighteen per cent of sample firms were manufacturing firms. Product profile of these firms include ceramics, sports wear, pipes, wine, plumbing material, and manufacture of auto components while 10 per cent of firms were manufactures of electrical and electronic products such as communication equipments, solar panels, and industrial scientific equipment. Ten per cent were constructions firms whereas 12 per cent were manufacturers of apparel.

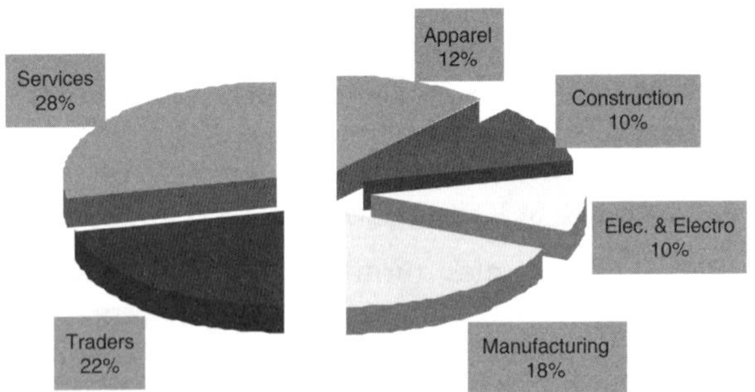

Figure 5.2   Industrial classification of sample firms

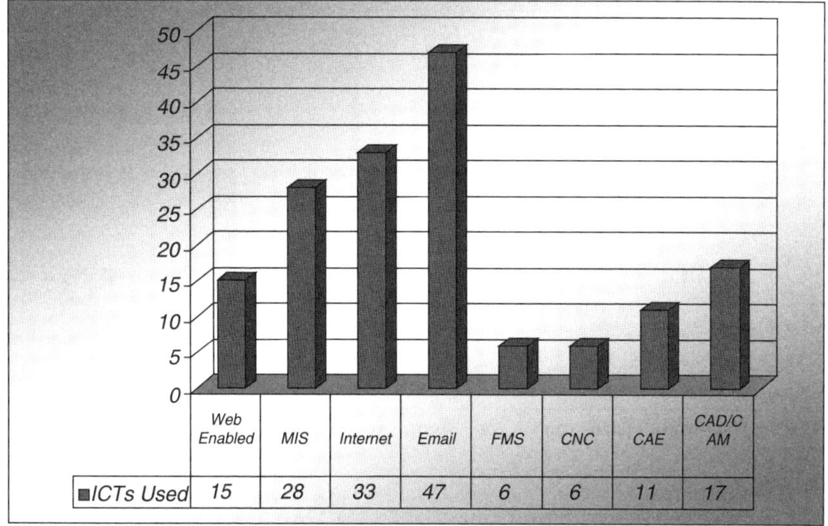

Figure 5.3   ICTs used by sample firms

6.1.2   Type of ICTs used

Types of ICT tools used by sample firms are presented in Figure 5.3. Email was found to be used by largest percentage (78.33 per cent) of firms followed by Internet (55.0 per cent). It may be noted that email was being used by more number of firms than Internet. This could be because of costly Internet infrastructure. Email is cheaper than the Internet. Hence firms might be using more email than Internet. Among production technologies, CAD/CAM was more popular than other tools. Low percentage (28.33 per cent) of its adoption is obvious because such tools are not very relevant for firms engaged in services and trading which are dominant sectors in the sample. FMS and CNC might have been adopted by electrical and electronic goods manufacturing firms that constitute only 10 per cent of the sample. It is worth mentioning that firms were using multiple tools hence percentage is not expected to add up to 100 per cent. It was found during the survey that firms that were using ICTs in production process were users of web enabled technologies, email, MIS, and Internet also.

6.1.3   Age of firms

Age distribution of firms is presented in Figure 5.4. It can be seen from the figure that roughly 12 per cent of sample firms were established before 1976. The oldest firm came into existence in 1959. The percentage of firms that came into being during 1976 to 1990 was 13.33 per cent. The growth of firms during early 1990s was much slower than mid 1990s.

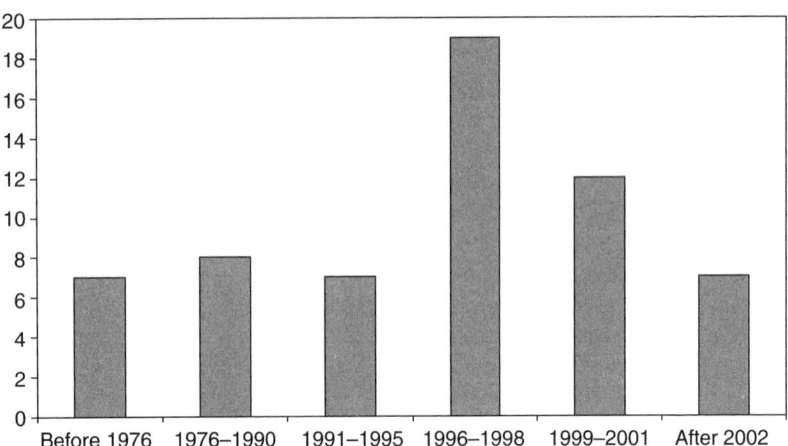

Figure 5.4   Age distribution of firms

Consequently 11.67 per cent of sample firms were established in 5 years during 1991 to 1995.

In general the growth of firms during the late 1990s was high in Jamaica. This is reflected by the fact that roughly 32 per cent of the sample firms were incorporated during 1996 to 1998. This pace of growth was slowed down in the next 3 years. Only 20 per cent of sample firms were established during 1999 to 2001. The growth of firms could not be sustained during latter period. Only 11.67 per cent of firms came into existence during 2002 to 2004. Newest firms came up during 2004.

6.1.4   Age of entrepreneurs

Distribution of firms by age of MDs is presented in Table 5.9. It can be seen from the table that roughly 48 per cent of MDs were in the age group of 46 to 55 years. A comparison of firms' age and the age of MDs suggests that

Table 5.9   Distribution of firms by age of MDs

Age group (years)	Frequency	Percentage
Up to 40	7	11.7
41–45	8	13.3
46–50	13	21.3
51–55	16	26.7
56–60	12	20.0
61 +	3	5.0
Total	59	98.3

majority of sample firms were not family business organizations. They might have been established by matured and experienced MDs due to SME promotional policies initiated by the Jamaican government. This is substantiated by the fact that only 20 per cent of firms are family-owned businesses.

6.1.5　Educational background of MDs

Firms distributed by MDs educational background is presented in Figure 5.5. One-third of MDs having no formal education or being studied up to high school suggests that they might be managing firms that are engaged in trading rather than manufacturing or service sector firms.

There seems to be a positive association between technology intensiveness of firms and the academic qualification of MDs. Apparently most of the electrical and electronic goods manufacturing firms were being managed by engineering degree holders. However this will be statistically tested in the following sub-sections.

6.1.6　Size of firms

Distribution of firms by size of employment is presented in Table 5.10. Minimum size of employment was one. There were two firms, i.e. bicycle parts trading and stationery selling that did not have any additional employees except owners. There was one firm dealing in construction

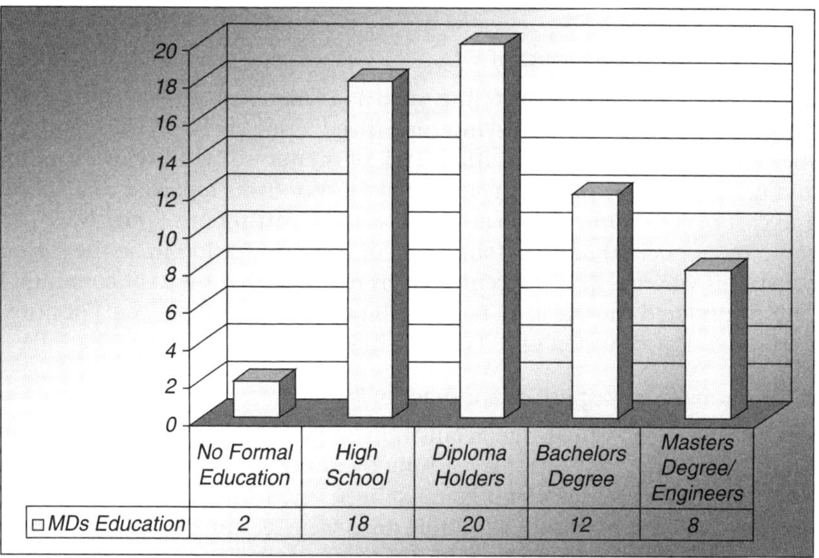

Figure 5.5　Distribution of firms by MDs education

Table 5.10   Distribution of firms by size of employment

Size of employment	Frequency	Percent
1–5	8	13.3
6–10	15	25.0
11–15	18	30.0
16–25	8	13.3
26–50	7	11.7
50 +	4	6.7
Total	60	100

business employed maximum number of workers, i.e. 75. The average employment size was 19 workers. Thirty per cent of sample firms employed workers between 11 and 15.

6.1.7   *International orientation*

Although sample firms were not dealing in international markets, many firms (21.7 per cent) had international partners. International collaboration was mainly for import of machinery and equipment. There was only one firm that collaborated for technical training. Apparel manufacturing firms had collaboration for import of garment designs. There was no firm that had financial collaboration with international firms. Major benefit of international orientation cited by firms was access to improved production technologies.

6.1.8   *Constraints in the use of ICTs*

Out of several potential constraints such as availability of Internet connection and subscription costs, Internet speed, usefulness of information, speed and cost of communication, lack of manpower with relevant skills and physical infrastructure, majority of MDs of firms reported that speed and cost of communication has been the major constraints. Roughly 77 per cent (76.7 per cent) of firms found cost of communication to be the major constraint whereas 23.3 per cent of firms reported that speed of communication hindered the use of ICTs. A small percentage (5.0 per cent) of firms attributed slow diffusion of ICTs to the Internet speed.

6.1.9   *Knowledge acquisition and mode of learning*

Several modes of knowledge acquisition such as formal training, on-job training by other institutions, learning by doing, learning from technical partners, and overseas trainings have been cited in the literature. Most prevalent modes of leaning in sample firms were learning by doing, on-job training by external sources, and formal training from professional training institutions. A majority of firms (51.7 per cent) found that learning by doing is most effective mode of knowledge acquisition while 38.3 per cent regarded on-job training provided by professional institutions as more effective. Merely 16.7 per cent of firms rated person with formal technical education are more suited. Response to effectiveness of various learning opportunities was a multi-response variable. Hence percentage of responding firms is not likely to add up to 100 per cent.

6.2   Statistical analysis

Statistical analysis has been performed in two steps. In first step composite score of ICT index has been generated and in second regression analysis has been used to assess the contribution of intensity of ICT adoption and other variables[13] to the performance of firms. As discussed earlier firms were using ICTs such as email, Internet, MIS, web enabled technologies, CNC, FMS, and CAD/CAM. Factor analysis was used to generate composite score of ICTs used by firms. Number of factors was controlled by counts rather than

Table 5.11   Descriptive statistics of determinants of ICT adoption [mean (STD)]

ICTs		MD_AGE	SIZE	SKILL	STO in million Jamaican $
Email	U	49.64 (7.34)	21.42 (19.94)	19.39 (18.82)	34.39 (46.44)
	NU	51.54 (10.24)	15.07 (12.82)	13.63 (11.75)	34.89 (52.05)
Internet	U	50.04 (9.26)	19.15 (17.31)	17.17 (15.74)	40.79 (52.30)
	NU	52.17 (6.12)	16.46 (17.66)	15.46 (18.23)	10.43 (13.77)
MIS	U	50.18 (8.06)	21.61 (19.71)	19.07 (18.45)	46.21 (57.19)
	NU	50.74 (9.39)	15.91 (14.62)	14.81 (13.87)	24.13 (37.08)
Web Enabled	U	48.20 (7.40)	32.13 (22.82)	27.40 (22.88)	50.59(60.52)
	NU	51.25 (9.06)	14.04 (12.27)	13.27 (11.52)	29.16 (43.22)
CNC	U	47.33 (7.37)	35.50 (15.90)	23.67 (16.01)	41.32 (13.68)
	NU	50.83 (8.84)	16.69 (16.49)	16.04 (16.15)	33.85 (51.04)
FMS	U	48.17 (4.22)	37.83 (22.13)	31.33 (19.94)	34.82 (17.59)
	NU	50.74 (9.07)	16.43 (15.46)	15.19 (15.06)	34.59 (50.99)
CAE	U	50.55 (4.08)	37.45 (23.94)	33.91 (23.14)	64.45 (65.68)
	NU	50.46 (9.49)	14.33 (12.13)	12.96 (11.29)	27.77 (41.66)
CAD/CAM	U	47.94 (8.54)	29.42 (22.55)	25.24 (22.12)	48.56 (59.07)
	NU	51.50 (8.67)	14.28 (12.60)	13.47 (11.85)	29.42 (43.68)
Total		50.47 (8.71)	18.57 (17.27)	16.80 (16.17)	34.61 (48.55)

Note: U→ Users and NU → non-Users.

---

13   Description of all the variables is presented in Appendix Table 5.2.

Eigenvalues. One factor was specified whose score represents the intensity of ICTs. Descriptive statistics of quantitative variables used in regression analysis were computed and are presented in Table 5.11.

Average size of operation of sample firms was 34.61 Million Jamaican Dollars. It can be seen from Table 5.11 that CAE using firms had the highest STO. In general firms using any type of ICT had higher STO than those that were not using a particular technology. For instance, average sales turnover of CAE using firms was 64.45 million J\$ whereas it was 27.77 million J\$ for non-CAE using firms. There was hardly any difference of STO in email using and non-using firms. Skill intensity and usage of ICTs follow the same pattern. Although size also follows the same pattern, the average employment in FMS rather than CAE using firms was highest. The pattern followed by MD_AGE in relation to usage of ICTs is very different. There is no significant difference of average between ICT using and non-using firms.

Subsequently data were analysed in regression analysis framework. The results are presented in Table 5.12. It can be seen from Table 5.12 that four different specifications of regression models have been used. This is because several independent variables were strongly correlated. The correlation matrix is presented in Appendix Table 5.1. In Model I all the variables (Appendix Table 5.2) were specified but MD_AGE and SIZE emerged significant because they are correlated with other independent variables. In Model II MD_AGE and SIZE were dropped and parameters were re-estimated. In the revised specification only skill emerged significant at 1 per cent. In Model III only foreign collaboration and communication

Table 5.12　Regression analysis

Dependent variables	Independent variable: LOGSTO			
	Model: I	Model: II	Model: III	Model: IV
Constant	0.572	2.377	2.371	2.618
MD_AGE	0.032*			
SIZE	0.035**			
FRN_COL	–0.141	0.117	0.984**	
SKILL	0.002	0.033***		
EXT_TRN	–0.297	–0.367		
COM_SPD	0.159	0.181	0.752*	
ICT_SCR	0.188	0.178		0.463***
R^2	0.485	0.353	0.132	0.126
F	4.583 ***	3.936***	4.172**	7.787***
N	41	41	57	55

Note: ***→ 1%, **→ 5 %, and *→ 10 % level of significance.

speed were included. In the last model we had to take composite ICT score as the only independent variables because it is highly correlated with all other independent variable (Appendix Table 5.1). It can be seen from Table 5.12 that ICT_SCR significantly influence performance of firms. Although R^2 of all four models is not very high, F-statistics is highly significant. This might be due to fact that many other variables that contribute to STO are not possible to include because of non-availability of data.

Although external training as mode of knowledge acquisition has not emerged significant when considered along with other variables, it is significant at the level of 5 per cent when specified alone in regression equation. This is because EXT_TRN follows the same pattern as ICT_SCR. The correlation between EXT_TRN and ICT_SCR is very high (0.603) with a significance level of 1 per cent. This finding is different than the perception that learning by doing is the best way of knowledge acquisition in SMEs.

Emergence of MD_AGE as a factor that influenced the performance of firms is also unique. There have been several studies (Earl, 1989; Lal, 1996) that find evidence of MDs' academic qualification influenced performance of firms. This study perhaps for the first time finds evidence of age of owners being played an important role in influencing performance. Age of MDs is a refection of strong knowledge base of MDs. The higher intensity of ICT adoption might be result of their formal training during which they might have been exposed to the potential benefits of ICTs. The argument is supported by the fact that external training which is provided by professional training institutions has been assigned more importance than other modes of learning modes by the MDs of sample firms.

The study also finds evidence to support the argument that more internationally oriented firms performed better than inward looking firms. This is not only in accordance with our hypothesis but support finding of other studies (Lal, 2004; Moreno, 1997). As discussed earlier most of the sample firms had technological collaboration rather than for marketing support. The results suggest that SMEs need to have access to better technology not only for international market but also to augment their performance in the domestic market. This is clearly due do globalization process which has affected SMEs as well.

Among the technological infrastructure variables, speed of communication has been found as significant factor that influenced performance of firms. Although speed of communication is external factor to the firms, firms could access to high speed communication network at a higher costs which may not be within reach of every SMEs. Given the size of Jamaican economy it may not be economically viable to encourage several private firms to enter into the communication sector. However, it is very necessary for government to provide broadband access to SMEs so that they remain competitive. The emergence of size as a critical determinant of performance is according to our expectation.

The other two factors that have influenced performance of firms are skill intensity of workforce and the intensity of ICTs adopted by firms. As discussed earlier skill intensity has been measured as number of persons using ICTs. The findings suggest that encouragement of ICT usage resulted in better performance of firms. Relationship between the intensity of ICT used and performance may be reinforcing each other suggesting that advanced ICTs using firms performed better than others and higher STO might have encouraged firms to adopt more advanced ICTs. The results suggest that SMEs need to be encouraged to use latest ICTs. These technologies should no longer be limited to non-production activities. Use of production technologies such as FMS, CNC, and CAD/CAM need to be encouraged. However, due care has to be taken for selection of appropriate production technologies as ICTs do not offer a panacea for all type firms. For instance CAD/CAM may be irrelevant for electronic metal sheet industry.

7   Summary and conclusions

The study is based on the primary data collected from 60 SMEs located in the Greater Kingston Metropolitan area. A semi-structured questionnaire was used to collect financial and historical data from sample firms. Data on technological profile and international orientation were also collected. ICT tools covered in the study include Internet, email, web enabled technologies, MIS, CAD/CAM, CAE, FMS, and CNC. Findings of the study suggest that most of the firms were using one or the other type of ICT. Internet was preferred technology used in the non-production processes whereas CAD/CAM was popular in production processes.

Factor analysis was used to generate a composite score of ICT intensity used by sample firms. Factor analysis included all the above mentioned ICT tools. Subsequently, neo-classical production function approach was used to identify the role of ICTs in augmenting performance of firm. The findings of the study suggest that the performance, represented by sales turnover was significantly influenced by intensity of ICTs used, skill intensity of workforce, foreign collaboration, age of owners/managing directors, size of firm, formal training as a mode of knowledge acquisition, and communication technological infrastructure represented by the speed of communication.

The study perhaps for the first time finds evidence of critical role being played the age of managing directors which is a proxy of their knowledge base. Although, age and academic qualification had positive association, the age has been able to capture the type of education provided to them and their knowledge base of new technologies. The study also finds evidence to support the argument that formal training of workers is preferred over learning by doing mode of knowledge acquisition. Emergence of managing directors' preference of formally trained workers in influencing size of operation as significant factor is a case in point. The other findings are in accordance with our hypotheses and support the findings of earlier studies (Lal, 2004; Earl, 1989; Doms *et al.*, 1997).

Globalization is characterized by three main components – the global-spatial proximity, the information and communication connectivity, and the integration of local economies into the global economy. Several authors have commented on the political economy and the social trans-formative aspects of globalization (Gibbon, 2002). However, the ability of local economies to participate in the global economy depends on the readiness of the local markets, a free trade barrier and their ability to connect with Northern markets through information and communication technology.

Technology is a development enabler through acceleration, driving, multiplier and innovator effects and holds the promise that SMEs can save on costs of production, become more flexible in terms of design, R&D, production, marketing, logistics, support services and the promotion of regional markets so that they become more efficient and more competitive in the global markets (Barnett, 1995). Exclusion from participating in the global economy is a major threat to Southern economies because of the complexity of markets and the specification of inputs and standard-ization of finished products, which make products of SMEs in the South uncompetitive in Northern markets. The opening of trade and capital flows also holds opportunities as well as threats because SMEs in the South may become unviable due to a flood of imported finished products.

Policy implications of the study are twofolds. One Jamaican government needs to improve accessibility of telecommunication network at a globally competitive rate. That in turn is expected in higher adoption of ICTs, which might influence performance of firms in the domestic market. Government also needs to provide marketing support to enable SMEs to venture into international markets. Greater integration of SMEs with the foreign market might contribution in export performance, which is very disappointing at present. Another important implication of the study is about the mode of knowledge acquisition. Government needs to strengthen the existing technical training institutions so that they can produce persons with appropriate skill needed by SMEs. Contrary to earlier realty that on job training is preferred mode of training is no longer a preferred option. We have not been able to assess the impact of financial policies of the Government on the performance of firms. A larger sample covering other areas of the country is needed to test the robustness of the findings of the study and to examine the spatial impact of the intensity of technology adoption and performance of firms.

Appendix Table 5.1   Correlation matrix

	LOGSTO	MD_AGE	SIZE	SKILL	EXT_TRN	COM_SPD	ICT_SCR
LOGSTO	1.00						
MD_AGE	0.259*	1.00					
SIZE	0.590**	0.085	1.00				
SKILL	0.553**	0.194	0.870**	1.00			
EXT_TRN	0.304*	−0.100	0.606**	0.590**	1.00		
COM_SPD	0.230	−0.105	0.330*	0.319*	0.485**	1.00	
ICT_SCR	0.355*	−0.147	0.496**	0.403**	0.603**	0.375**	1.00
FOR_COL	0.280*	−0.058	0.311*	0.211	0.407**	0.010	0.468**

Note: **→ 1% and *→ 5 % level of significance.

Appendix Table 5.2   Variables used in the analysis

Variables	Remarks
LOGSTO	Natural log of sales turnover which was measured in million J$
MD_AGE	Age of managing director
SIZE	Size of workforce
FRN_COL	Foreign Collaboration: Measured on binary scale
SKILL	Number of persons using ICTs
EXT_TRN	Effectiveness of external learning: Measured on a binary scale
COM_SPD	Speed of communication a constraint? Measured on a binary scale
ICT_SCR	ICT score: Computed from factor analysis

6
Globalization and the Adoption of ICTs in Nigerian SMEs[14]

1 Introduction

The recent wave of globalization spurred largely by rapid advances and diffusion of Information and Telecommunication Technologies (ICTs) has led to irreversible changes in the world economy. Globalization now encompasses not only the financial markets, but also the whole gamut of the world's social, political, economic and cultural milieu. This era of globalization commenced in the 1980s although it already began to rise significantly in the preceding decade. The process of globalization is rapid, complex and full of uncertainties and expectations. It is obvious that only those countries that invest in science and technology (S&T) are able to integrate themselves better in the global system and to effectively respond to the challenges posed by globalization.

Consequent upon this new era, the African continent in general and Nigeria in particular, are faced with new challenges in the process of socio-economic development. One dimension of this is how Nigeria can further integrate with the global economy through enhanced trade, and still cope with the prospect of unfettered economic growth and greater openness. This challenge is enormous especially given the current low level of penetration of processed and manufactured goods as well as services from Nigeria into the world trading system. Between 1999 and 2004 the value of exports from Nigeria increased from NGN 1,543,707 million (US\$ 16,816m) to NGN 3,257,536.5 million (US\$ 24,401.02m), although manufactured goods accounted for only 1.0 per cent and 2.1 per cent of the exports in 1999 and 2003 respectively (Federal Office of Statistics, 2004; CBN, 2004).[15] Exports especially industrial goods from Nigeria and other developing

---

14 Abridged version of the chapter has been accepted for publication in Science Technology and Society.

15 The Official exchange rate was NGN 91.80 and NGN 133.50 to one US Dollar in 1999 and 2004 respectively (CBN 2003 and 2004).

nations are confronted with some competitive forces such as rivalry of competitors within its industry, threats of new entrants, threats of substitutes, the bargaining power of customers and the bargaining power of suppliers. The awesome capabilities of the various ICTs to confront some of these forces have been proved in a number of countries such as the United States, Brazil, Guatemala, and Tunisia (UNCTAD 2003, 2004a). The available data show that international trade in ICT goods and services has grown in recent years at a faster rate than total international trade and that it remains robust (UNCTAD, 2004a). The impact of ICTs on the performance and competitiveness is achieved through increased information flows, which result in knowledge transfer as well as improved organization. The application of ICTs in business ranges from Computer-Aided Design (CAD), Computer-Aided Engineering (CAE) and Computer-Aided Manufacturing (CAM) to Intranet, Internet and Electronic Commerce (E-Commerce) among others. Writing in the 2004 UNCTAD Commerce and Development Report, Kofi Annan noted that 'E-Commerce is one of the most visible examples of the way in which ICTs can contribute to economic growth. It helps countries into the global economy. It allows business and entrepreneurs to become more competitive. And it provided jobs, thereby creating wealth' (UNCTAD, 2004a).

International competitiveness in trade continues unabated in spite of the growing number of supra-national organizations such as the European Union (EU), North American Free Trade Zone, Southern Common Market (in Spanish *Mercado Común del Sur* (MERCOSUR), Economic Community of West African States (ECOWAS), and the Southern African Development Community (SADC). While African countries are widely regarded as being non-competitive, there are a few success stories. Some garment-producing firms in Mauritius have gained prominence as exporters even to the EU and USA markets since the early 1980s. UNCTAD (2004b) notes that since the USA's Africa Growth and Opportunity Act (AGOA) came into effect in 2000, many export oriented garment factories have been set up in Namibia, Mozambique, Swaziland, Lesotho as well as Mauritius. The report went further to say that Lesotho became the largest African apparel exporter to the USA in 2003, with about 10,000 newly created jobs. Oyeyinka-Oyelaran (1997a) has commented on the emerging Nnewi industrial cluster in Nigeria as another success story, even though the manufactures especially automotive spare parts are exported to other West African countries.

Improving business competitiveness requires a variety of systemic transformation, including the availability of efficient infrastructure and services, technological changes, new organizational processes, adoption of good practices, training, mobilization of underutilized resources, creation and segmentation of markets, product and process certification and efficiency in inputs utilization. A study of Small and Medium Enterprises (SMEs) in OECD on the challenges of globalization in OECD reveals that, internal dynamics of firms and environmental support are key factors which could either facilitate or inhibit export. SMEs that are active players in the global economy make special efforts to search for diversified growth by pursuing innovation-based production, and open-minded management capable of engaging the appropriate specialized resources such as ICTs. The enabling environment includes effective consulting, funding and logistical resources to support exports. Some of the key internal factors inhibiting the globalization of SMEs are lack of experience on the part of the management, inadequate resources and an excessive risk perception. With regards to the environment, poor national information networks with weak or inadequate international connection, poor regional resources and support programmes are some of the prominent examples (Julien *et al.*, 1993).

Much as it is acknowledged that globalization and competitiveness of SMEs and Large Scale Enterprises (LSEs) are driven by similar forces, the pre-eminence of ICTs in this regard is unassailable. A critical question for developing countries is whether the adoption of ICTs can be an authentic vehicle for solving the problems of economic development and low level participation in international trade. In this chapter we critically examine the general response to the apparent threats to Nigerian SMEs posed by the globalization of production and markets. The two main policy research questions examined in the chapter are: What strategies have proved useful in terms of enabling Nigerian SMEs to become more competitive through the use of ICTs? And how can ICTs be used to facilitate the participation of Nigerian SMEs in national and international supply chains?

This remaining part of the chapter is organized into seven sections comprising an introductory section, wherein the research problem and major policy questions are articulated.

Section 2 starts with the conceptual aspect of the study with ample review of the issues of globalization, competitiveness, adoption and diffusion of technology. The factors influencing technology adoption are expected to provide important background to appreciating the reasons why firms are competitive locally and internationally. The hypothesis guiding the study is also highlighted in this section. Section 3 presents some of the relevant literature on the research theme. In this regard an overview of the Nigerian economy is presented, followed by discussions on the nation's SMEs including their performance, contribution to the economy, capacity utilization, employment, exports and institutional support for them. The section is concluded with a brief discussion on ICT and globalization.

Section 4 describes the evolution of structural and organizational changes in SMEs in the framework of global commodity-chain and production networks using largely international examples. However, this section also focuses on the recent developments in Nigeria's ICT sector. This section shows that Nigeria is a late starter in ICT development and adoption as it were. Section 5 provides detailed description of the research methodology we adopted for the study. It describes the sampling and data gathering procedures and the framework of data analysis.

The presentation of the empirical study findings and discussion is the focus of the next section. In addition to the usual discussions on the sample firms, size distribution and age, affiliation to international firms and technical staff, we also present the results of the ordered PROBIT regression model. Section seven contains discussion on selected case studies of firms that have adopted some ICTs and are very competitive in their lines of business. In the last section, we present the summary of major findings, policy recommendations and some concluding remarks.

2   Conceptual framework

We would prefer to discuss several aspects of globalization before a conceptual framework for this study is delineated. In spite of widespread interest in its emergence and impact, there is overwhelming evidence to show that unanimity has not been reached on the definition of the concept of globalization. Definitions of the complex concept vary and in some cases, the definitions are restricted to selected aspects. In the view of Julien (2001) globalization of the economy is distinct from the traditional trend towards internationalizations. This is because the globalization is manifested by the broadening of exchanges of goods and services of all types, particularly in developing nations. He asserts that a significant proportion of these exchanges involve affiliates or branches of multinational firms (intra-firm exchanges) which with their strong development over the preceding decades, have accelerated 'delocalized production'. Many of these exchanges he further noted are the outcomes of the formation of new alliances or agreement, between businesses as observed in the automotive, fine chemistry or computer industries (Julien, 2001).

For lawyers, globalization concerns the threatened changes in legal status of states and their citizens. For environmentalists, it is the changes in the world's climate and other biological systems. And for information technology experts, globalization means the worldwide spread and integration of information system (Lee, 2003). From the economic perspective Lee (2003) defines globalization as the 'process of changing the nature of human interaction across a wide range of spheres including the economic, political, social, technological and environment.... the process of change can be described as globalizing in the sense that boundaries of various kinds are becoming eroded. This erosion is taking place along three dimensions: spatial, temporal and cognitive (Lee, 2003). IMF (1997) defines globalization as the economic interdependence of countries worldwide through the increasing volume and variety of cross-border transactions in goods and services and of international capital flows, and also through the more rapid and widespread diffusion of technology'. Oyeyinka (2004), contends that globalization is the deep on-going socio-economic changes and especially associated with, and employed to explain the financial market and technological competitiveness. He says it is also related to the increasing integration of national economies with that of the rest of the world.

The obvious popularity of this phenomenon known as globalization is attributable to two main factors. The first is its scale and its speed and the way that technology in particular ICTs, but also in transportation – is changing the world for all of us. Advances in ICTs and transportation mean that large segments of national economics are much more exposed to international trade and capital flows than hitherto (Lal, 2004). Second, it is the most current economic process that is widely accepted as changing the international environment and turning the entire world into a global village (Ajayi, 2004).

Nevertheless, there are some concerns about this phenomenon, namely: Globalization represents a new form of colonial exploitation; National governments have been replaced as colonial aggressors by the global institutions and the transnational corporations; The central driving forces are free-market capitalism, liberalization and Third World debt; The outcome will be the creation of a 'global economic elite' whose interests transcend national boundaries and a marginalized and subjugated 'world proletariat' (Kirkbride *et al.*, 2001).

A study on globalization and China concludes that while wealth has been created, overall income inequality has increased. But most regions in the country including lero open areas, have exhibited a drastic reduction in poverty counts (Wei, 2002). It is in this context that Joseph Stigliz commented that 'globalization is neither good nor bad'.

It has the power to do enormous good, and for countries of East Asia, who have embraced globalization under their own terms, at their own pace, it has been an enormous benefit, in spite of the setback of 1997 crisis. But in much of the world, it has not brought comparable benefits. For many, it seems closer to an unmitigated disaster.

... the benefits of globalization have been less than its advocate claim, the price paid has been greater, as the environment has been destroyed, as political processes have been corrupted, and the rapid pace of change has not allowed countries time for cultural adaptation. The crises that have brought in their wake massive unemployment have, in turn, been followed by longer term problems of social dissolution ... (Joseph Stiglitz, 2001).

In spite of the divergent opinions on the definitions and implications of globalization, the features of the current global economy can be represented by: rapid advances in new technologies; expanding spatial scope for the business activities of private enterprises including financial institutions;

integration of markets across national borders; a higher degree of uniformity in policy and institutional environments that establish the rules of the game for economic actions and interactions of private agents located in different parts of the world; the volume of international trade has increased dramatically; and private capital flows from the developed to developing economies have also risen significantly since the mid-1980s, with the bulk of these flows going to emerging market economies (Court and Yanagihara, 1998; Ajayi, 2004; Kose *et al.*, 2004).

The foregoing scenario is quite significant for Nigerian SMEs. This is because it is only those SMEs that can carve a niche for themselves in the production process, that are competitive through the adoption of very relevant marketing strategies and are advanced in the development and adoption of latest technologies and the ideas are likely to benefit most from globalization. In other words, it is presupposed that the proportion of SMEs acting as global players will depend on their ability to muster the required resources and adjust to this rising competition and interdependence. On the one hand SMEs will have to compete with big business (multinationals and trans-nationals) and on the other compete with themselves.

A number of authors have examined the concept of competitiveness from different theoretical perspectives. Competitiveness when applied at the enterprise level, relates to profits or market shares; and when applied at the country level it relates to both national income and international trade performance, particularly in relation to specific industrial sectors that are important in terms of, for example, employment, productivity and growth potential. Perhaps, more important from the conceptual perspective is that competitiveness is based on Schumpeterian logic that sees the nature of capitalist development as a sequence of innovative investments associated with dynamic imperfect competition and productivity gains, and that sees a major role for public policy in facilitating productivity-increasing investment (UNCTAD, 2004b).

Linkage between capital accumulation, technological progress and structural change constitute the basis for rapid and sustained productivity growth, rising living standards and successful integration into the international economy. It is generally not easy to totally and correctly accommodate micro-level analysis of firm and industry competitiveness with micro-analysis of national development and prosperity. Alanen (1996) argues that the competitiveness can be examined separately at firm, sectoral or the whole economy level depending on the ultimate utility of such studies.

The definition of competitiveness as a concept is dynamic depending on the aspect of the economy concerned. At the firm level, competitiveness refers to a company's ability to attract or increase market share. Global competitiveness can be defined as the capability of a company to generate more wealth than its competitors in world markets.

Competitiveness is decomposable into four factors namely unit cost, productivity, profit margin and exchange rate. There are however other intervening social and institutional variables in competitiveness. The continuous improvement and innovation are key determinants of international competitiveness, and also that competitive sectors in a national economy are not discrete entities, but exist in form of clusters of interconnected industries. Towards this end, firms can achieve competitive advantage through cost leadership, offering products with superior value that justify a premium price, or focusing on a specific buyer group, segment or geographical market. Offerings are defined as all products or services that come under customer choice in a given buying situation (Alanen, 1996). Consequently, a strategy of competitiveness is essentially concerned with the positioning of the firm's outputs, not of inputs. That is, what matters to the prospective buyer in the decision-making process is price of the good or service and not its costs (Table 6.1).

Table 6.1   **Summary of factors impinging on firm competitiveness**

Dimension of competition	Sub-dimensions	Determining factors
Costs (proxy for price)	Magnitude of related net cost increase	Type and extent of environmental externality (e.g. Pollution intensify, climate change, loss of biodiversity)
	Competitive significance	Technological aspects (e.g. investment potential, innovation potential) Ability to absorb (e.g. cost structure, profit margins) Ability to pass on cost to customers (e.g. price elasticity of demand, prices of competing product)
Differentiation	Feasibility of competition based on differentiation of products Differentiative importance of good practices (e.g. environmental characteristics)	Nature of product (e.g. price elasticity of demand, prices of competing productions) Customer's good practices concern (e.g. eco-labelling, destination of exports, sensitivity of issue)
Business alliances	Development of inter-business alliances and information systems	Increase in strategic business virtual alliances with suppliers, customers etc

Source: Alanen (1996).

From Table 6.1, we note that at the firm level, competitiveness is a function of productivity, environmental performance, quality performance and market performance. Different basic strategies are required to meet these four challenges and ICTs especially the Internet and CAD/CAM, have been deployed appropriately in some parts of the world as critical aspects of technological capability.

Ability to harness and use scientific knowledge to solve socio-economic, cultural, engineering and related problems is one perspective of technological capability. Technological capability has been defined as the great variety of knowledge and skills which need to 'acquire, assimilate, use, adapt, change and create technology' (Ernst, *et al.*, 1994). According to these authors, technological capability incorporates non-engineering and technical know-how. It embraces organizational know-how, knowledge of behavioural patterns of workers, suppliers, and customers. A firm's technological capabilities in the area of e-commerce, e-business, engineering and manufacturing for meeting competitiveness challenges and improvement of efficiency and good-keeping practices can be described as highly essential in this era of globalization (see also Adeboye, 1995).

Under globalization, technology transfer is expected to assume greater importance. Bell and Pavit (1993) developed a model to describe the process of international technology transfer and accumulation in developing countries by explaining how the knowledge and skills are learnt from production process fed into the technological capabilities of firms. In this regard, some emphasis was placed on the role of investment in skills development, training and special schemes for acquiring experience from all available sources. The potential for adopting various ICTs by firms in order to develop and/or acquire these capabilities is both incontrovertibly high and real.

Dosi *et al.* (1988) and Kim (1997) have ably demonstrated that in all economies, the firm (enterprise) is central to all the key activities associated with technological innovation. The underlying precept is that successful firms react appropriately to the forces of technology push and the attraction of demand pull. Appropriate reaction entrails the development and adoption of relevant technologies (Adeoti, 2005). Several factors working either singly and or synergistically determine the adoption of a new technology such as the Internet or CAD. A firm's consumer behaviour and or the producer behaviour represent the drivers of the adoption process. These drivers are basically competitive forces faced by firms. Examples of such forces are: bargaining power of customers and suppliers; rivalry of competitors; threat of new entrants; and threat of substitutes.

Technology diffusion is essentially a multi-stage process which begins with acquisition of the technology and finally, the installation, utilization and assimilation of the technology. The assimilation process is perhaps the most critical, since it involves adaptation to the local environment. The failure of many imported technology in Nigeria has been traced to failure in the process of assimilation.

The adoption of ICTs creates strategic information systems (SIS), information system that appropriately supports or shapes the competitive position and strategies of the firm. The strategic role of SIS involves their utilization for the development of products, services and capabilities that confer on an enterprise major advantages over the competitive forces it faces in the global market place. Consequently, and as vividly demonstrated by Rosegger (1989); Bessant (1999); Lundvall and Battese (2000); Adeoti (2002, 2005), the proactive firm would adopt technology (ICT in this instance) to meet with the challenges posed by the competitive forces. The major strategic approaches and how the adoption of ICTs can be used to foster competitiveness are presented in Table 6.2.

Experience of successful organizations reveals that extensive and meaningful managerial and end user involvement are the key ingredient of high

Table 6.2   Potential roles of ICTs in achieving competitive advantages

Strategic approach	Role of ICTs
Cost minimization	– To reduce cost of business processes through enhanced efficiency and productivity – Reduction of transaction costs including costs of customers or suppliers
Differentiation of products and services	– New ICT features for differentiation – Can be used to reduce the differentiation advantages of competitors – Focus products and services at selected market niches
Innovation	– For creating new products and services that may also include ICT components – For developing unique new markets or market niches – Making radical changes to business processes aimed at reducing costs, improving quality, efficiency, or customer service, or shorten time to market (e.g. CAD, CAM)
Growth promotion	– Managing local, regional and global business expansion. – Diversification and integration into other products and services
Development of alliances	– Creation of virtual organizations of business partners – Development of inter-enterprise information systems linked by the Internet and extranets that support strategies business relationships with customers, suppliers, sub-contractor and others – Strong supply chain management

Source: O'Brien, James (2003).

quality performance of the information system. In addition to skills, successful e-businesses have re-engineered their organizational structures and roles, as well as their business processes, so as to ensure they are agile, customer-focused and value-driven. Figure 6.1 presents a model developed by Neilson *et al.* (2002) to describe both pre-e-businesses and e-organizations and this model is as germane today as it was in 2002.

The macro-economic, social and institutional environments within which production and marketing firms operate have been identified as one of the key determinants of technology diffusion in developing countries.

Figure 6.1   A model of pre-organizations and e-organizations

Pre-e-Organization	e-Organization
Organization Structure	
★ Hierarchical	★ Centerless, networked
★ Command-and Control	★ Flexible structure that is easily modified
Leadership	
★ Selected 'stars' step above	★ Everyone is a Leader
★ Leaders set the agenda	★ Leaders create environment for success
★ Leaders force change	★ Leaders create capacity for change
People and Culture	
★ Long-term rewards	★ 'Own your own career' mentality
★ Vertical decision making	★ Delegated authority
★ Individuals and small teams are rewarded	★ Collaboration expected and rewarded
Coherence	
★ Hard-wire into processes	★ Embedded Vision in individuals
★ Internal relevance	★ Impact projected externally
Knowledge	
★ Focused on Internal processes	★ Focused on Customers
★ Individualistic	★ Institutional
Alliances	
★ Complement Current gaps	★ Create new value and outsource uncompetitive sources
★ Ally with distant partners	★ Ally with Competitors, Customers, and Suppliers
Governance	
★ Internally focused	★ Internal and external focus
★ Top-down	★ Distributed

Source: Neilson *et al.* (2002).

The schematic framework presented in Figure 6.2 shows the key drivers of SMEs in Nigeria. The relationships between the various boxes in this figure are in some cases recursive and deterministic. For example, physical infrastructure such as roads, rail, sea ports, water, telecommunication and electricity are expected to increase allocative efficiency of both marketing and production. However, if such infrastructures are poor and inefficient, these benefits would become unrealizable as it is the current situation in the country. In view of the foregoing discussion we therefore hypothesize that the adoption of ICTs by firms is largely influenced by the size, history, international orientation as well as management and structural characteristics of firms.

3   Literature review

3.1   An overview of the Nigerian economy

After a long period of weak performance, the economic situation in Nigeria has somewhat improved in the last few years. According to the most recent available data, the real gross domestic product (GDP), which grew at the rate of 0.40 per cent in 1999, leapt to 3.50, 10.23 and 6.5 per cent in 2002, 2003 and 2004 respectively. The real growth rate of the oil sector was almost 24 per cent in 2003 compared to 4.4 per cent for the non-oil sector

Figure 6.2   Key drivers of SMEs development in Nigeria

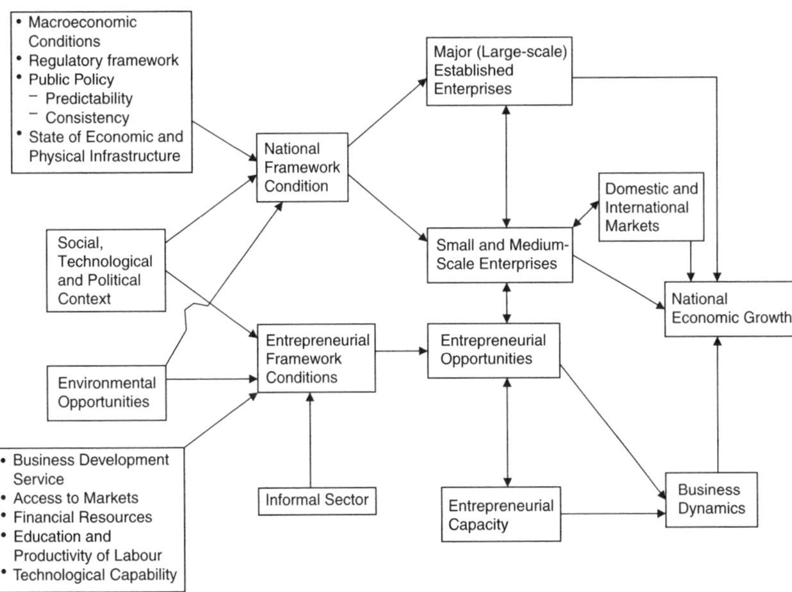

Source: J. Alos (2001) Presentation to the SME Working Group Meeting at the 8th Nigeria Economic Summit, Abuja.

Table 6.3   Percentage distribution of Nigeria's GDP by economic activities

Activity	2000	2001	2002	2003
Agriculture	26.29	34.04	34.85	**30.74**
Crop production	22.04	28.52	29.18	25.72
Livestock	2.57	3.30	3.39	2.91
Forestry	0.49	0.59	0.61	0.58
Fishing	1.19	1.60	1.67	1.53
Mining & quarrying	48.29	35.75	38.42	**39.58**
Coal mining	0.00	0.00	0.00	0.00
Crude petroleum & natural gas	48.19	35.75	33.29	39.46
Metal ores	0.00	0.00	0.00	0.00
Quarrying & other mining	0.10	0.13	0.13	0.12
Manufacturing	3.71	4.25	4.39	**4.14**
Oil refining	0.10	0.29	0.26	0.25
Cement	0.06	0.07	0.07	0.06
Other manufacturing	3.55	3.89	4.06	3.83
Utilities	0.05	0.05	0.06	**0.05**
Electricity	0.03	0.03	0.04	0.03
Water	0.02	0.02	0.02	0.02
Building & construction	0.67	0.87	0.89	0.85
Wholesale & retail trade	11.62	13.72	14.30	13.27
Hotel & restaurants	0.14	0.17	0.18	0.16
Transportation	2.85	3.08	3.31	**3.23**
Road Transport	2.56	2.77	2.97	2.96
Rail Transport & Pipelines	0.0	0.00	0.00	0.00
Water Transport	0.02	0.02	0.02	0.01
Air Transport	0.04	0.04	0.05	0.04
Transport Services	0.23	0.25	0.27	0.22
Post telecommunication	0.02	0.03	0.04	**0.04**
Telecommunication	0.01	0.02	0.02	0.03
Post	0.01	0.01	0.02	0.01
Banks & financial institutions	0.96	1.06	1.47	**1.16**
Financial institutions	0.93	1.12	1.42	1.13
Insurance	0.03	0.04	0.05	0.03
Real estate	3.64	3.97	4.05	4.18
Business services (not health or education)	0.06	0.07	0.07	0.06
Public administration	0.60	1.34	1.37	1.07
Education	0.14	0.30	0.28	0.24
Health	0.04	0.08	0.07	0.06
Private non profit organizations	0.00	0.00	0.00	0.00
Other services	0.91	1.11	1.23	1.13
Broadcasting	0.01	0.02	0.02	0.02
GDP at basic prices	100.00	100.00	100.00	100.00

Source: Federal Office of Statistics (2004).

during the same period. Table 6.3 shows that the primary sector of the economy (agriculture and mining and quarrying) continues to dominate total output. In 2000, the sector contributed 74.58 per cent falling slightly to 70.3 per cent in 2003. Inflation averaged 12 per cent in 2003, compared to about 13 per cent in 2002. Due to the generally higher world oil prices in 2003 compared to 2002, the overall balance of payments exhibited a much healthier improvement (Federal Office of Statistics, 2004). The sector's average contribution to the GDP between 2000 and 2003 was 70.82 per cent. Since the primary sector remains the main driver of the recovery process, the situation may be rather fortuitous and, hence unsustainable since the performance of this sector is exogenously determined.

Table 6.4 depicts the structure of Nigeria's foreign trade. Within the context of this study we observe that 95 per cent of total exports is accounted for by oil and gas, while imports concentrate more on consumer goods and machinery and equipment. Imports of intermediate goods such as raw materials averaged about 26 per cent between 1999 and 2003. On the average, oil constituted about 98 per cent of total primary exports for the period under review (Appendix Table 6.1).

While the manufacturing sector has the potential to create employment and wealth, stagnation continues to characterize the sector. Hence, these two potentials have remained largely unattained. The activity mix in the sector is also limited, dominated by import-dependent processes and factors. Manufacturing capacity, which was very high up to the mid-1980, has been struggling to recover since the introduction of the Structural Adjustment Programme (SAP) in 1986. The capacity utilization rate, which was 42.7 per cent and 44.3 per cent in 2001 and 2002 respectively managed to rise to 45.0 per cent in 2004 (FRN, 2004; CBN, 2004). The real growth rate of this sector rose favourably from 3.44 per cent in 2000 to 10.07 per cent in 2002 before falling to 5.7 per cent in the following year. Additional information on the sector is contained in Table 6.5.

Availability of an efficient physical infrastructure in an economy is very critical to sustainable enterprise development. Under this condition, networking among enterprises is greatly facilitated especially though the use of telecommunication services, while transaction costs are lowered. In Nigeria the state of the physical infrastructure is quite pathetic. For example in 2003, the total electric power installed capacity was 6,472 Mega Watts (MW) while available capacity was a mere 2,060 (MW) less than 32 per cent (Federal Office of Statistics, 2004). The result is severe and frequent power outages compelled firms to purchase standby electric power generating sets. Similarly, water supply is inadequate and in some locations just not available. Other infrastructure facilities and services such as roads and railways are also in poor conditions (Olokesusi, 1987; NISER, 2003).

It is no wonder then that the nation's economy has been performing very well below optimality. A study of business activities conducted by the

Table 6.4   Imports and exports during 1999–2003 (NGN million)

Description	1999	2000	2001	2002	2003
Total imports	406,962	591,325.6	885,114.0	1,054,075.6	1,923,098.8
%	100.0	100.0	100.0	100.0	100.0
Consumer goods	175,400.6	237,121.6	330,074.0	402,656.9	694,238.4
% of total imports	43.1	40.1	38.6	38.2	36.1
Intermediate goods	115,984.2	153,744.5	261,993.7	259,302.6	734,623.7
% of total imports	28.5	26.0	29.6	24.6	25.8
Equipment & machinery	115,577.2	200,459.4	281,466.3	392,116.1	732,700.6
% of total imports	28.4	33.9	31.8	37.2	38.1
Total exports	1,559,300	2,745,479.6	2,007,127.02	2,239,381..9	3,109,288.4
%	100.0	100.0	100.0	100.0	100.0
Primary goods	1,543,707	2,739,478	2,001,105.6	2,105,023.7	3,043,993.3
% of total export	99.0	99.6	99.7	94.0	97.6
Other manufactured goods	15,593	11,001.9	6,021.4	134,363.2	65,295.1
% of total export	1.0	0.4	0.3	6.0	2.1

Source: Federal Office of Statistics (2004).

Table 6.5   Manufacturing sector performance

Manufacturing	1999	2000	2001	2002	2003	Mean
Real growth rate of manufacturing (percent)	3.44	3.44	6.99	10.07	5.66	5.92
Manufacturing sector value added	214,224.25	238,501.48	285,844.48	343,999.07	402,978.93	297,109.64
(NGN million):						
Compensation of employees	8,628.93	9,605.87	11,512.88	14,301.86	16,753.92	8,809.91
Consumption of fixed capital	4,123.48	4,524.14	4,964.28	6,164.11	7,225.05	5,400.21
Operating surplus	138,194.12	153,907.00	184,915.28	229,721.23	269,103.71	195,168.27
Value added at factor cost	150,946.52	168,037.02	201,392.69	250,187.19	293,082.69	212,729.22
Indirect taxes	63,297.75	70,464.46	84,451.79	93,811	109,896.29	84,384.26
Subsides	0.00	0.00	0.00	0.00	0.00	0.00
Value added at current producer price	214,244.25	238,501.48	285,844.48	343,999.07	402,978.93	297,113.64
Intermediate consumption	272,768.32	300,348.57	340,760.54	408,055.15	487,333.77	361,853.27
Gross output	487,012.61	538,850.05	626,605.03	752,054.21	890,312.70	658,966.92

Source: Federal Office of Statistics (2004).

Federal Office of Statistics (2004) revealed that security of lives and property topped the list of constraints (80 per cent) perceived to be serious followed by power (electricity) supply (73.79 per cent), corruption (71.72 per cent), unpredictability of currency exchange rate (66.21 per cent) and delay and uncertainty in the implementation of government policy (64.14 per cent) (Table 6.6).

In addition to all these constraints are low technological capacity of support institutions, low technological capability of personnel, poor management of resources, low patronage of domestic research and development institutions because of strong linkages between local enterprises and foreign partners, strong competition with imported machinery and products. These set of constraints impact negatively on SMEs due to absence of effective business development services (Osoba, 1988; Akerele, 2003). Despite the numerous constraints highlighted above, there is appreciable optimism and confidence in the economy among the existing businesses at least in the medium-term based on the optimistic plans of the sample firms (see Table 6.7).

3.2 Review of Nigerian SMEs

3.2.1 History of Nigerian SMEs

In Nigeria, there is no universally acceptable definition of SMEs as it had varied overtime and from organization to organization. In 2001, the National Council of Industry (NCI) put the capital investment band for SMEs at between NGN 150–NGN 200 million inclusive (excluding land but including working capital) and the working force band at between 11 and 300 inclusive. On the other hand, the National Association of Small and Medium-Scale Enterprises (NASME) defines a small-scale enterprise as a business with the number of staff employed by the enterprise less than 50 and with an annual turn-over of NGN 100 million. NASME went on to define a medium-scale enterprise as a business with the number of staff employed by the enterprise less than 100 and with an annual turnover of NGN 500 million. The Central Bank of Nigeria (CBN) defines an SME as any enterprise with a maximum asset base of NGN 200 million excluding land and working capital, with the number of staff employed by the enterprise expected to be not less than 10 and not more than 300 people. Table 6.8 provides a summary of SME definitions in the country. For the purpose of this study, enterprises with a working force band of between 10 and 300 people inclusive, and with an asset, excluding land but including working capital of less than NGN 200 million were regarded as SMEs.

Historically, SMEs in Nigeria have existed since independence in 1960. Since the period of independence, Nigeria has had series of studies, seminars and workshops, each of which extolled the excellence, importance and need to facilitate the establishment and sustenance of SMEs. All the National 4-Year Development Plans from the 1962–68 to 1981–85, have

Table 6.6   Constraints in business development in Nigeria

Constraint factors to development and growth	Percentage distribution of constraint perception				
	Not a problem	Insignificant problem	Moderate problem	Serious problem	Total
Security of lives and property	3.45	5.52	11.03	80.00	100
Uncertainty about government police					
Delay and uncertainty in implementation of govt. policy	4.83	9.66	21.38	64.14	100
Budget	6.90	15.17	26.90	51.03	100
Reversals of policies	8.97	12.41	30.34	48.28	100
Unpredictability of exchange rate	4.14	11.03	18.62	66.21	100
Poor infrastructure					
Power supply	4.83	6.90	14.48	73.79	100
Fuel and other petrochemical supply	3.45	14.48	22.76	59.31	100
Roads	5.52	9.66	28.97	55.86	100
Ports services	13.10	12.41	33.79	40.69	100
Customs clearance procedures	7.59	17.24	33.79	41.38	100
Water supply	17.93	17.93	16.55	47.59	100
Corruption	6.21	7.59	14.48	71.72	100
Finance					
Access to long-term credit	9.66	14.48	19.31	56.55	100
Interest rate on bank credit	4.83	8.28	30.34	56.55	100
High level of taxation	4.14	10.34	29.66	54.46	100
Multiple taxation	4.14	10.34	18.62	62.07	100
High tariff rates on input	8.97	15.86	20.69	54.48	100
Social and political unrest, including labour strikes	5.52	11.72	22.07	60.69	100
Access to land for investment and expansion	27.59	20.00	17.24	35.17	100
Others (specify)	5.52	4.14	2.07	88.28	100

Source: Federal Office of Statistics (2004).

Table 6.7 Proposed investment in business establishments during 2004–2007

Legal status → Investment year	Limited	Joint stock company	Corporation	Others	Total
2004					
Amount	358,366,000,014	2,342,250,000	77,850,013	156,117,359,000	516,903,459,027
Per cent	69.33	0.45	0.02	30.2	100
2005					
Amount	669,971,620,000	1,420,300,003	112,561,850,000	375,001,600,000	1,158,955,370,003
Per cent	57.81	0.12	9.71	32.36	100
2006					
Amount	1,325,836,600.00	1,242,400,000	40,960,250,000	690,018,000,000	2,058,057,250,000
Per cent	64.42	0.06	1.99	33.53	100
2007					
Amount	2,356,461,595,016	1,002,043,750	154,171,793,763	1,222,280,646,501	3,733,916,079,030
Per cent	63.11	0.03	4.13	32,73	100
Total					
Amount	4,710,635,815,221	6,006,993,754	307,771,743,788	2,443,417,605,597	7,467,832,158,360
Per cent	63.08	0.08	4.12	32.72	100

Note: The amount is in NGN million.
Source: Federal Office of Statistics (2004).

Table 6.8 Definition of SME by Nigerian institutions

Institution	Asset value (N'm)			Annual turn-over (N'm)			No. of employees		
	MSE	SSE	ME	MSE	SSE	ME	MSE	SSE	ME
Fed. Min. of Industry	<200	<50	n.a	n.a	n.a	n.a	<300	<100	<10
Central Bank	<150	<1	n.a	<150	<1	n.a	<100	<50	n.a
SMIEIS	<200	n.a	n.a	–	–	–	<300	–	–
NERFUND	n.a	<10	n.a	n.a	n.a	n.a	n.a	n.a	n.a
NASSI	n.a	<40	<1	n.a	<40	n.a	n.a	3–35	n.a
NASME	<150	<50	<1	<500	<100	<10	<100	<50	<10

Source: World Bank (2001).

each laid strong emphasis on strategies of government-led industrialization hinged on import-substitution. With the initiation of the SAP in 1986, the state downgraded its active involvement in industrialization by a process of commercialization and privatization. Emphasis therefore shifted from large-scale industries to SMEs, which have viable potentials for developing domestic linkages for rapid and sustainable industrial development. Greater leanings were placed on the organized private sector (OPS) to spearhead subsequent industrialization programmes (NIPC, 2002). The sector was further actively encouraged by additional incentives. These incentives were directed at solving or at least alleviating the enormous problems encountered by industrialists in the country and thereby afforded them greater leeway towards increasing their contribution to the national economy.

3.2.2 *Performance/contribution of SMEs to the national economy*

SMEs currently represent about 90 per cent of the industrial sector in terms of number of enterprises, however they contribute a meager 1 per cent of gross domestic product (NIPC, 2002). This is insignificant when compared to countries like Indonesia, Thailand and India where SMEs contribute almost 40 per cent of GDP. Whilst SMEs constitute an important part of the business landscape in any country, they are faced with significant challenges that compromise their ability to function and to contribute optimally to the economy. To have an idea of how large the SME sector is in Nigeria, the Corporate Affairs Commission in Abuja estimates that about 90 per cent of all Nigerian businesses in 2001 employed less than 50 people (NIPC, 2002). Similarly, a study conducted by the International Finance Corporation (IFC) during the same period estimated that 96 per cent of all businesses in Nigeria are SMEs (compared to 53 per cent in USA and 65 per cent in the EU, with SMEs in both places accounting for over 50 per cent of their respective country's GDP) (NIPC, 2002). This shows that given necessary support SMEs constitute important players in the development process of Nigerian economy and has proved to be one of the most viable sectors with economic growth potentials.

An insight into the investment activities and earnings of SMEs can be gained by examining the findings of the Nigerian Institute of Social and Economic Research (NISER, 1998–2003) and Akinbinu (2003). NISER has been surveying the business conditions, experience and expectations of the operators of the Nigerian manufacturing sector since about a decade. The survey has largely included the SMEs in the major industrial clusters (Lagos, Kano, Port Harcourt, Asaba-Onitsha-Nnewi and Kaduna-Jos axes) in Nigeria. A study by Akinbinu (2003), among other issues, examined the performance of SMEs in Western Nigeria. Some findings of these studies are appraised and supported by other relevant study findings.

3.2.3   SMEs earnings and investment

Findings from the NISER survey which is presented in Table 6.9 reveal that, over the period 1998–2003, less than 40.0 per cent of the respondents indicated a rise in their earnings except in 2000 when 47.5 per cent of them did so. This implies that majority of them have either experienced a fall in earnings or their earnings remain unchanged. Similarly Akinbinu (2003) finds that SMEs who sustained losses rose from 17.0 per cent in 1996 to about 46.0 per cent in 2000, while those who perceived their profit level to be good fell from about 10.0 per cent in 1996 to about 5.0 per cent in 2000. In the case of investment, it can be seen from Table 6.9 that, apart from 2000, less than 40.0 per cent of the responding manufacturers revealed that their investment levels increased over the period 1998–2003. Thus, the table shows that, over the period, the proportions of respondents who indicated that their investment level remained unchanged ranged between 35.0 and 53.0 per cent. Only between 14.0 and about 31.0 per cent reported that their investment level rose.

Table 6.9   Distribution of firms by ratings of earnings and investment

Year	Ratings of earnings				Ratings of investment			
	Rise	Fall	Remain constant	Total	Rise	Fall	Remain constant	Total
1998	29.6	40.8	29.6	100	22.2	37.9	39.9	100
1999	37	20.4	42.6	100	32.4	14.9	52.7	100
2000	47.5	31.4	21.1	100	40.8	24.2	35	100
2001	34.1	32.9	33	100	31.4	27	41.6	100
2002	37.6	29.8	32.6	100	37.4	27.2	35.4	100
2003	29.6	27	43.4	100	29.1	31.1	39.8	100

Source: NISER Survey of Business Conditions, 1998–2003.

3.2.4   Capacity utilization, employment and output

It can be seen from Table 6.10 that capacity utilization in the small-scale industry is low. It has not reached 45.0 per cent during the period 1990 to 1999. Capacity utilization in the medium-scale industry which ranged between 50.0 and 57.0 per cent in 1990–95 dropped to less than 40.0 per cent during 1998–99. In the same vein, capacity utilization in large-scale industry which was below 50 per cent in the 1990–95 period dropped down as far as about 20–27 per cent in the late 1990s. The findings of the NISER survey of business conditions confirmed low capacity utilization in the SMEs.

Table 6.11 revealed that the responding manufacturers declared that they are operating below the installed capacity (CBN, 2004; Federal Office of Statistics, 2004). Given these low capacity utilization rate, it is expected that employment and output will be low. The NISER survey of business

Table 6.10   Capacity utilization by size of firms (percent)

Year	Small	Medium	Large	Annual mean
1990	42.5	55	48.5	48.6
1991	41.2	56.6	46.4	48.1
1992	42.8	53.4	45.6	47.2
1993	37	50.2	44	43.4
1994	35	49.7	44.8	42.7
1995	32.3	51.1	43.6	41.8
1996	30	41.1	39.3	36.8
1997	30	41.5	39.9	37.2
1998	28.6	34.5	20.3	27.8
1999	28.8	36.4	27.2	30.8
Mean (1990–99)	34.8	47	40	40.4

Sources: Oyeyinka (1997a) and Akinbinu (2003).

Table 6.11   Distribution of firms by ratings of capacity utilization

Year	Ratings of capacity utilization			
	High	Low	Very Low	Total
1997	22	78	–	100
1998	19.3	80.7	–	100
1999	29.2	55.3	15.5	100
2000	26.9	62.3	10.8	100
2001	38	54.2	7.8	100
2002	40.2	50.8	9	100

Source: NISER Survey of Business Conditions, 1998–2003.

Table 6.12 **Percentage distribution of firms by rating of output and employment**

Employment year	Ratings of output				Ratings of employment			
	Rise	Fall	Remain constant	Total	Rise	Fall	Remain constant	Total
1998	29.7	33.3	39	100	17.2	39.3	43.5	100
1999	39.3	14.2	46.5	100	27.5	18.2	54.3	100
2000	26.8	53.4	19.8	100	31.8	33	35.2	100
2001	23.9	49.5	26.6	100	28.9	29.8	41.3	100
2002	23.6	58	18.4	100	35.6	29.7	34.7	100
2003	37.5	25.7	42.7	100	24.8	32.5	42.7	100

Source: NISER Survey of Business Conditions, 1998–2003.

conditions reported the ratings of these performance indicators by the responding manufacturers.

In 1998, majority of the manufacturers indicated that their output either fell or remained the same, while only about 30.0 per cent of them experienced an increase (Table 6.12). Further, a relatively high proportion of them reported that output remained constant, while less than 40.0 per cent of them stated that it increased in 1999. In the period 2000–02, a relatively high proportion of the respondents declared that output fell, while majority of them revealed that output either remained constant or declined in 2003. In the case of employment, it can be seen from the table that, over the period 1998–2003, less than 40.0 per cent of the responding manufacturers indicated rise in employment levels while majority of them found it same or declining. These analyses have shown that the performance of the SMEs has not been impressive over time. The policy environment has not been favourable enough to promote output and employment in the industrial sector.

At the aggregate level, the contribution of the manufacturing sector to GDP has ranged between 5 and 6 per cent since 1980 and up till now (CBN, various years). In the manufacturing sector, the share of small-scale industry (SSI) in total output has been less than 15 per cent, while the contribution of large-scale enterprises is substantial (Table 6.13). Therefore, the constraints militating against the performance of SSI should be addressed.

3.2.5   *Export by Nigerian SMEs*

Over the period 1997–2002, less that 20 per cent of the responding manufacturers indicated that they were exporters. Thus, overwhelming majority of them stated that they did not export (Table 6.14). In 1997 and 1998, among those that are exporting / less than 30 per cent of them stated that

Table 6.13   Contributions of small and large scale industries to manufacturing sector output

Year	Share of small-scale industry in manufacturing output	Share of large-scale industry in manufacturing output	Growth rate of SSI output
1981	14.78	85.22	–
1982	14.54	85.46	11.07
1983	13.62	85.38	–33.89
1984	14.33	85.67	–6.56
1985	14.33	85.67	19.86
1986	14.33	85.67	–3.89
1987	14.33	85.67	5.1
1988	14.30	85.70	12.54
1989	14.33	85.67	1.92
1990	14.32	85.68	7.62
1991	14.31	85.69	9.3
1992	14.33	85.67	–4.84
1993	14.30	85.70	–4.14
1994	14.39	85.61	–0.89
1995	14.33	85.67	–5.50
1999	14.33	85.67	3.50
2000	14.31	85.69	3.49
2001	14.30	85.70	4.11

Source: Underlying Data Obtained from the Federal Office of Statistics, National Accounts Statistics.

Table 6.14   Distribution of firms by export orientation

Year	Responses		
	Yes	No	Total
1997	17.2	82.8	100
1998	14	86	100
1999	15.6	84.4	100
2000	16.2	83.8	100
2001	18.1	81.9	100
2002	16.4	83.6	100

Source: NISER Survey of Business Conditions, 1998–2003.

their volume of export was increasing, while the rest reported that it was either decreasing or remained constant.

It can be seen from Table 6.15 that the proportion of those who revealed that their export volume was rising increased to over 40 per cent in 1999 and 2001. It can be inferred from this analysis that only few of the

Table 6.15   Percentage distribution of firms by trends of export

Year	Trends of Export			
	Increasing	Decreasing	Stable	Total
1997	22.1	36.3	41.6	100
1998	23.5	56.7	19.8	100
1999	44.2	25.3	30.5	100
2000	32.2	36.7	31.1	100
2001	43.4	39.1	17.5	100

Source: NISER Survey of Business Conditions, 1998–2003.

responding manufacturers exported during the period and for majority of them who exported the volume of export was not increasing. Therefore, the contribution of SMEs to foreign exchange earning of the economy will be low. This finding is similar to the findings on the same issue by Adeoti (2002).

3.2.6   Appraisal of institutional support to SMEs in Nigeria

The SMEs, given their basic characteristics of small capital investments, small sizes, low profit margin and little management/staff; cannot afford all the technical and support services that they need for successful operations. However, realizing that SMEs hold the greatest prospects of growth for the Nigerian economy, the government over the years has begun to address the constraints that impede their growth by putting in place arrays of programmes and policies that would provide an enabling operating environment (see Figure 6.2). These support programmes and policies as documented in NIPC (2002) and Akinbinu (2003) include – financial services, export promotion, technical services for innovation, training and investment promotions. In other words, much of these interventions aimed at addressing institutional and market failures and constraints to SMEs survival and growth, through financial and technical assistance programmes.

The financial support programmes included incentives in the form of subsidized targeted credit scheme administered through the Development Finance Institutions (DFIs) and sectoral interventions through selective credit control contained in the credit guidelines issued by the Central Bank of Nigeria (CBN). Others were technical and training support and classification of SMEs as preferred sector.

The first systematic effort at supporting SMEs was the setting up of the Industrial Development Centres (IDC) in each state starting with Owerri, Osogbo and Zaria. These were complemented with the establishment of DFIs such as the Nigerian Industrial Development Bank (NIDB), Nigerian Bank for Commerce and Industry (NBCI), National Economic Recovery

Fund (NERFUND) and the Nigerian Export and Import Bank (NEXIM). Subsequently, the specialized and subsidized targeted credit schemes such as NERFUND and the World Bank SME I and SME II loans were established. These credit schemes provided interest rate subsidy and included the participation of the commercial and merchant banks in administering the loans.

Following the collapse of the oil boom in the 1980, government renewed its policy employing export incentive package to promote the diversification and expansion of exports. These incentives included Manufacture-In-Bond (MIB) scheme, Import Duty Drawback scheme, Export Development and Expansion Fund.

Government at the time undertook SAP in mid-1986 with the expectation that new entrepreneurial spirits and initiatives would be unleashed to make SMEs flourish. Much emphasis was placed on SMEs to look inward, develop and increase their raw materials supply and boost output and export trade. In 1988 (2 years after the commencement of SAP), government recognized the promotion of SMEs as a more appropriate strategy for achieving the nation's target for industrial self-reliance, employment generation and poverty alleviation and hence the introduction of a new industrial policy (Adeoye, 2002). The introduction of this policy was an attempt to mobilize substantial loanable funds for lower-scale industrial development through development banks, the People's Bank and World Bank projects.

Moreover, in order to meet the credit needs of this sector, the government has over time set up some specialized financial institutions including the Nigerian Industrial Development Bank (NIDB) and the Nigerian Bank for Commerce and Industry (NBCI). Additionally, government approved the establishment of Community Banks by the private sector. In addition, the Federal Government has merged the People's Bank with NIDB to create the Bank for Industry with a take-off fund of 50 billion. Also, the banks constituting the Banks Committee have agreed to set aside 10.0 per cent of their pre-tax annual profits for equity participation in small industrial ventures with private entrepreneurs. The scheme is called the Small and Medium Enterprises Equity Investment Scheme (SMEEIS). This 10 per cent profit before tax shall be invested as equity investment in SMEs as the banking industry's contribution to the Federal Government's efforts towards stimulating economic growth, developing local technology and generating employment. Funding to be provided under the scheme shall be in the form of equity investment in eligible industries. This will reduce the burden of interest and other financial charges expected under normal bank lending, as well as provide financial, advisory, technical and managerial support from the banking industry. As at November, 2004, the SMIEIS fund was about NGN 28.8 billion, while only NGN 9.33 billion representing 32.3 per cent had been invested in 173 enterprises throughout the country (*The Guardian*, 2004).

3.3   ICT and globalization

ICT and globalization have as earlier observed in this study, combined to create a new economic and social landscape. Indeed, they have initiated fundamental changes in the functional processes and procedural framework of enterprises and economies. That the comity of nations and development institutions attach tremendous importance to them is borne out by the numerous fundamental and high profile initiatives being taken at the international level. These initiatives have the singular objective of promoting the development and adoption of ICT. Some of these initiatives are the World Summit on the Information Society (WSIS), the G8 DOT Force (Digital Opportunity Task Force), the United Nations ICT Task Force in addition to the various national ICT programmes. These initiatives attain further significance in view of the catalytic roles, which ICTs could play in development generally and in particular, the attainment of the UN Millennium Development Goals (MDGs) (UNCTAD, 2003).

The World Bank (2004) asserts that there is growing integration of societies and economies. Between 1990 and 2000 the number of the poor people living on less than US$ 1 a day fell by about 137 million. The gross private capital flows as a proportion of the global GDP rose from 10.1 per cent in 1990 to 20.8 per cent in 2002. Gross foreign direct investment (FDI) as a proportion of global GDP similarly rose from 2.7 per cent in 1990 to 6.0 per cent in 2002. Trade in goods as a share of GDP, which was 32.5 per cent in 1990 picked up to 40.3 per cent in 2002 (World Bank, 2004).

Also, due to globalization, there has been an increase in foreign direct investment (FDI) flows to Africa in general and Nigeria in particular. In 1990, FDI inflow to African nations was around US$ 2.5 billion, lagging far behind official development assistance and other official flows. By 2002, this figure rose to US$ 11.7 billion and then rising by 28 per cent to US$ 15 billion out of the total global FDI of US$ 172 billion in 2003. The 2004 edition of the UN Conference on Trade and Development (UNCTAD) annual report – *World Development*, predicts that FDI flows to Africa will further increase in subsequent years.

It is however instructive to note that FDI flows to Africa are skewed disproportionately towards extractive industries particularly oil. Typical examples are Algeria, Angola, Chad, Equatorial Guinea, Libya, Nigeria and Sudan. In both 2002 and 2003, seven out of the top ten recipients of FDIs were oil producers. With the possible exception of Algeria, which is now diversifying the investment inflows into non-oil sectors such as steel, chemicals, pharmaceuticals and telecommunications, such investments have had only limited development benefits in the other oil producing countries.

Julien (2001) identified the several elements as characterizing the worldwide spread of globalization or the previous growth of international exchanges. The elements are: the accelerated lowering of customs barriers, opening new export possibilities; increasing competition fostered on national SMEs by the preceding element; accentuation of competition by diversification and rapid replacement of goods and services, thereby limiting the lifespan of national products; increased competition due to the presence of multinational in the manufacturing and services industries; capability of the multinationals to take over vibrant SMEs or to conclude a range of alliances with them or engage them as subcontractors; increased access by SMEs to diverse materials and immaterial resources especially information, or at least compel them to become systematically attentive to change in the world; and possible reduction in government subsidy to SMEs because of new competition rules enthroned by supra-national bodies.

Using these characteristics as background, Julien (2001) was able to establish that most (70 per cent) of the manufacturing SMEs he surveyed in Quebec, Canada were relatively sensitive to heightened global competition especially the Canada – U.S. Free Trade Agreement while 30 per cent of the responding enterprises were well-informed on this free trade agreement, about 70 per cent were found to have made recent investment and are involved in research and development (R&D) so as to set their product apart. This group also felt that they must do all things possible to enhance their competitiveness to deal with the increase in international pressure, or seek niches or novel ways of doing business (Julien, Joyal and Deshales, 1993). Another study in Quebec by Julien and Carriere (1993), confirmed the above results. The use of at least one computerized or 'leading-edge' technology increased from about 13 per cent in 1986 to over 50 per cent in 1992. However, the use of ICT in manufacturing systems by SMEs lagged behind that of large businesses.

4 Evolution of structural and organizational changes in SMEs

This section examines how ICTs might improve the overall competitiveness of SMEs, the near ICT revolution in Nigeria and how firms are changing their behaviours and networking so as to take advantage of ICT in trade competitiveness.

Developments in the field of ICTs have resulted in the emergence of a variety of innovations, which tend to be radically altering the nature of business enterprises, particularly the SMEs. The wide spread replacement of under production, specialized, single-purpose, fixed equipment by computer aided design and engineering capabilities, robots, automatic handling and transporting devices, flexible manufacturing system (FMS), computer aided/integrated manufacturing (CAM/CIM), just-in-time (JIT) techniques, manufacturing resource planning (MRP) and telematics have permitted firms to be more efficient and competitive (Alcorta, 1992). As opposed to the old system of production technology, the new system of production characterized by ICT leads to qualitatively diverse command and control structures, work organization patterns and competitive strategies. Emphasis now is on human capital development and acquisition of a multi-skilled, creative, knowledgeable, trustworthy and responsible work force. Further, producers' relationship with suppliers, customers and competitors, which was previously at a distance, is now conducted on collaborative links. Thus, competition, which had been on cost minimization (or price cutting) basis is now on the basis of continuous innovation and prompt market response. ICT introduces strategic element into the production system by equipping the decision-making unit with the capacity to alter the product mix either by developing new product(s) or by improving the quality of the existing ones (Alcorta, 1992).

As far as the role of ICTs in promoting competitiveness of firms is concerned, ICTs promote operational efficiency of firms by lessening processing and delivery time (and hence, minimize processing and delivery costs), transaction and inventory costs, and minimizes material wastage. It can provide strategic benefits to firms especially in the area of competition by way of facilitating production of new products or controlling distribution channels, thereby resulting in lower prices, improvements in design and products quality and hence, enhance competitiveness (Cane, 1992; Akinboyo, 2004).

According to the literature, the adoption of ICTs is expected to enhance competitiveness of firms by transforming them from a condition of low level of output to a condition of mass production. It thus facilitates participation in global trade by raising information content of economic activities worldwide, in terms of access to information on prices, qualities standards, users, producers and users of commodities. It also provides opportunities to interact with various producers and users as need arise. Marcelle (2004), in his study of technological learning by firms in developing countries contend that it is internal competencies, such as the ability to manage strategic change and develop coherent systems that enable those firms to effectively navigate technological frontiers, the network of global suppliers and weak national innovation systems.

The World Summit on the Information Society (WSIS) Plan of Action emphasizes the role of national e-strategies as key instruments for the advancement of the information society. It also calls for the use of ICTs by SMEs to foster innovation, achieve gains in productivity, reduce transaction costs and combat poverty. The E-commerce and Development Report prepared by UNCTAD (2004b) contends that ICT usage usually increases with the size of firms, although SMEs were found to have the greatest potential for productivity gains through e-business. The report also revealed that in Asia and Africa, many urban-based SMEs have connected to the Internet and actively use it for communicating with suppliers and customers.

A survey of SMEs in Chile, Colombia, Costa Rica, Mexico and Venezuela conducted by UNCTAD and Foundation for sustainable development among SMEs (FUNDES), shows that the availability of personal computers (PCs), and the Internet were high among urban-based firms, and there was no significant difference between small and medium sized firms as regards basic access to and use of the Internet. But the sampled SMEs did not frequently use ICTs for more complex tasks, especially automating and integrating business process. E-commerce was found to be still rare and small firms used more e-market places, whereas medium-sized firms used company website (of third parties or their own) for selling online. The most active users of ICTs were service firms, followed by trade and manufacturing enterprises respectively. This finding is similar to the outcomes of studies in other developing regions. The plausible explanation is that functions like marketing and selling services online require basic Internet access and website presence, and less system integration is required for supply and value chain management as is the case in manufacturing.

UNCTAD (2003), reports that African business-to-business (B2B) e-commerce was expected to grow from US$ 0.5 billion in 2002 to US$ 0.9 billion in 2003, with South Africa accounting for 80–85 per cent of these values. Online auctions for coffee are now held annually in Brazil, Guatemala and Nicaragua a situation that has led to the integration of traditional marketing and improvements in export market for coffee.

Under normal circumstances, SMEs procure international information with respect to both inputs and outputs to determine the evolution of markets and products and therefore of competition, both upstream and downstream. Thus, their export success depends very much on their commercial competitive and technological monitoring (Verhoeven, 1988; Ali and Swiercz, 1991). Studies have also shown that some SMEs form network between and within their countries of location (international SME networks) so as to export together in a value chain or linked to firms, which are already major players in international markets. In many countries, industrial districts or clusters provide the opportunities for networking. Under this circumstance, certain firms regarded as 'pivot firms' provide leadership and the others are directly linked to them either as importers of raw materials or in the area of marketing and distribution to other countries in addition to facilitating new markets (Sengenberger *et al.*, 1990; Julien *et al.*, 1993). Similarly, some Quebec-based SMEs have been found to form international alliances with firms in Francophone African countries, and their export success was a function of their ability to develop information and thereby exchange knowledge and skills to the advantage of the alliance members (Marcotte and Julien, 1994).

In Nigeria, the use of ICT has taken a positive dimension. Many private firms have invested in ICTs, while there are business centres offering ICT services. Firms have been acquiring personal computers (both desk- and lap-tops), email and Internet equipment and services. A study by Adeoti (2004) provides an insight into the ICT investment activities of the Nigerian manufacturing firms. The study reveals that investment in ICT hardware in the Nigerian manufacturing industry ranges from those on computers, electronic sensors, electronic inventory monitoring units, and computer aided manufacturing equipment. Analysis of survey results in this study shows that, prior to 1990, the use of ICTs was not considered important, thus, most firms started using it only 3 years ago. It was discovered that, prior to 1990, the use of ICTs in the Nigerian manufacturing sector to a large was limited to affiliates of multinational enterprises. It is revealed in this study that all responding manufacturing firms have acquired some simple forms of ICTs. It was reported that majority of these firms have not acquired ICT hardware that have direct applications in production system adopted. Thus, only 8.1 per cent of them have acquired computer-aided equipment in their manufacturing activities. It was found that 18.9 and 17.6 per cent of the responding firms have electronic inventory monitoring units and sensors respectively, while 12.8 per cent of them have Digital Cameras to monitor their production system. Hence, based on the foregoing analysis, it can be said that there has not been an extensive application of ICT in the Nigerian manufacturing industry.

In general, in Nigeria, the application of ICTs in economic activities is far below expectations. According to Bamkole (2005) 67 per cent of 201 sample SMEs, indicated their desire for assistance in the areas of automation and information technology. The factors responsible for this poor performance include the high level of poverty and low level of development in the areas of infrastructure and human capital in the economy and high cost of ICTs. The prevailing high level of poverty has led to low level of effective demand, and hence low level of investment in all areas of business activities including acquisition of ICT equipment. The introduction of ICTs has therefore not been in concordance with development of supporting infrastructure, education and training of users (Olayiwola and Ogundiran, 2004).

One of the key priorities of the Obasanjo administration since mid-1999 has, therefore, been how to reinvigorate the economy and reposition the decayed physical infrastructure. Towards this end, the government produced a National Information Technology Policy (NITP) in March, 2001. The vision statement of NITP recognizes the importance of ICT in achieving global competitiveness and says *inter alia* .... 'to make Nigeria an IT capable country in Africa and a key player in the information society by the Year 2005, using IT as the engine for sustainable development and global competitiveness'. Chapters 8 and 14 enunciate the objectives and strategies for mainstreaming ICTs in trade and commerce and global competitiveness respectively (Federal Ministry of Science and Technology, 2001).

The National Information Technology Development Agency (NITDA) was established in the Federal Ministry of Science and Technology to midwife the new policy. NITDA has taken some significant steps especially the initiation of the e-governance project at the federal level since 2003. It also engaged a number of relevant stakeholders for the purpose of proposing policy recommendations to the government on IT issues. The stakeholders include: Nigeria Computer Society; Nigeria Internet Group; Internet Service Providers Association of Nigeria; Nigeria Society of Engineers; Nigeria Software Development Institute; and Institute of Software Practitioners of Nigeria.

The government was supposed to provide US$ 100 million takeoff grant to NITDA. However, less than 20 per cent of this amount was released to it, thus, reducing its impact on the sector. Meanwhile, the government has established the Engineering Materials Development Institute (EMDI) in Akure to undertake R&D in the area of engineering materials and engineering materials infrastructure for the production of various types, sections and sizes of machines and equipment. Presently, EMDI is researching in the areas of CAD/CAM as well as virtual manufacturing. Virtual manufacturing entails the use of modeling and simulation techniques to visualize (with the computer) the effects of design and process parameters, prior to actual manufacturing (*The Punch*, 2005).

Since the Digital Mobile License (DML) auction in January, 2001, Nigeria has left the group of countries with the lowest teledensity in the world. Prior to the DMI auction there were less than 500,000 fixed lines. According to Ernst Ndukwe, the Executive Vice-Chairman of the Nigerian Communication Commission (NCC) the telecommunication regulatory agency, the active subscribers in mobile networks was more than 7 million lines in the last quarter of 2004. In 2003, there were about 3 million lines up from 1.5 million, although fixed line subscription has not increased as rapidly as the mobile lines, the number actually rose by 100 per cent between 2001 and 2004. As at March, 2005, there were 12.5 million mobile lines in the country and the Nigerian telecommunication sector experienced a robust real growth rate of 47 per cent and 43 per cent in 2002 and 2003 respectively.

The Federal Minister of Communication revealed in April, 2005, that investment in the ICT sector since 2001 has exceeded US$ 4 billion with several thousands of jobs created. MTN which is leading the pack has about 1,800 workers in its employment. Of this number, 65 are South African, 23 are expatriates from other countries and 1,722 (95.5 per cent) are Nigerians (Aragba-Ekpore, 2005, p.80; IT and Telecom Digest, 2005).

The GSM operators (MTN, Glo Mobile, V-Mobile and M-tel) and the fixed wireless telephone service providers e.g. Mobitel and EMIS have in recent times introduced a bouquet of value-added content and business products. One of these is the v-active and v-office introduced by V-mobile. The new products run on Short Message Service (SMS), Circuit Switched Data (CSD) and General Packet Radio Service (GPRS) are supported by platforms like

Multi-Media Messaging Service and Wireless Application Protocol (WAP). V-active is designed for accessing multiple entertainment and information options, while v-office provides Internet browsing, fax, email and remote network access for enterprises and business executives.

To give further boost to the sector the Nigerian Information Practitioners in the USA plan to establish 40 ICT parks in 39 universities across the country between 2004 and 2006. Tagged the AfriHUB project, the first two parks commenced operations in October, 2004, at both the Nsukka and Enugu campuses of the University of Nigeria, AfriHUB's second initiative into bring to the country reliable C-band broadband very small Aperture Terminal services designed and priced to penetrate the SMEs market many of which currently rely on Ku-band. The institutional investors in AfriHUB project include Skyterra Communications of the US, Zinox Technologies Nigeria, Rainbownet Nigeria Limited, some fixed wireless access and Internet Service Providers (ISPs) (*The Punch*, 2004).

President Olusegun Obasanjo in February, 2005, inaugurated the Nigeria Software Development Initiative (NSDI) with a view to repositioning the country and derive alternative source of income. NSDI is charged with the responsibility of developing realistic strategies for the development and promotion of software products and services in the country. In March 2005, a solid foundation was laid for the formation of a non-governmental organization, the Nigerian Internet Registration Association (NIRA) for the purpose of administering the Nigerian country code Top Level Domain (cc TLD). If properly managed, it is highly speculated that Nigeria can and should actually save and generate revenue from locally managing the ccTLD. In addition to the financial gains, the pride of locally managing such a resource is significant, while more people and enterprises would register domain names at lower rates than hitherto (IT & Telecom Digest, 2005).

5 Data and study methodology

The study was designed to cover as many SMEs as possible for three reasons. First, manufacturing and services have become more knowledge intensive across a wide range of industries such as forestry, textile, furniture, ICTs and clothing firms among others. Knowledge in this regard include R&D, product design, process engineering, management and knowledge about market investment opportunities and the skills and capabilities needed to undertake changes in products and processes, create networks and sustain partnering activity. Second, competition now depends more on the capability of firms to innovate within the context of survival in a globalized environment (Mytelka and Tesfachew, 1998; Mytelka, 1998). Third is the necessity to ascertain the extent of ICT adoption and diffusion since the unprecedented changes began in 2001. The sample firms were identified with the aid of Membership Directories obtained from the National

Association of Small and Medium-Scale Enterprises (NASME) in Aba, Ibadan, Lagos and Nnewi. In addition, we obtained the list of exporters from the Lagos head office of the Manufacturers Association of Nigeria (MAN). From these directories, a sampling frame was generated for each of the four locations after cleaning for duplications. To obtain the maximum cooperation of the firms, a letter of introduction was written to each firm's management informing same of the study objectives and the need for their cooperation. Two copies of the research instrument were attached to each of these letters before they were manually lodged followed by eventual retrieval by trained research assistants after several repeat visits and telephone calls. The principal investigator carried out the case studies.

Within the context of ICT and globalization, the survey questionnaire was designed to elicit information on: general profile of the sample enterprises; international orientation; ICT adoption; consequences of using ICTs; and competitiveness.

In the sample firms, the target respondents were the managing directors and where this was impossible, attention shifted to the plant and or operations manager. The selected firms are located in two of the country's major industrial zones namely Lagos-Otta-Ibadan and Aba-Nnewi-Onitsha. The other two zones are Port-Harcourt; and Kano-Kaduna. A stratified random sampling procedure was adopted to select a sample of 31 enterprises in Lagos and 24 in each of Aba, Ibadan and Nnewi. The sample for Lagos is slightly higher because it has more enterprises than any of the other three locations. In all, 103 enterprises were sampled by well-trained research assistants in the four cities. However, the analysable research sample consists of 67 firms (representing 65 per cent of total), the sectoral distribution of which is shown in Table 6.16.

Table 6.16   Sectoral distribution of sample firms

S/N	Sectors	Number	Percentage
1.	Paper and paper products (PPPs)	6	8.96
2.	Rubber and rubber products (RRPs)	7	10.45
3.	Chemicals and chemical products (CCPs)	14	20.90
4.	Non-ferrous metals and mineral products (NFMMPs)	1	1.49
5.	Iron and steel basic industries products (ISBIPs)	16	23.88
6.	Textile and textile products (TTPs)	3	4.48
7.	Leather and leather products (LLPs)	1	1.49
8.	Food, beverages and tobacco products (FBTPs)	8	11.94
9.	Wood and wood products (WWPs)	3	4.48
10.	Agricultural and agro-allied products (AAAPs)	5	7.46
11.	Personal computer sales and services (PCSSs)	3	4.48
	Total	67	100.0

The enterprises sampled include those in the industrial, commerce and service sectors. The enterprises selected for the study in each of the sectors were carefully chosen with a view to reflecting the true position of their respective sectors in terms of sales, turnover, employment structure, and ICT adoption among others. Published and unpublished materials were also sourced from libraries, Federal Office of Statistics (now National Bureau of Statistics), government departments and the Internet in order to complement the primary data. Three enterprises were also examined as case studies and they provided detailed information about their operations and benefits associated with the adoption of ICTs.

6 Empirical findings and discussion

6.1 Profile of the sample firms

Sample firms were grouped into three categories, namely: Non-adopters of ICTs; Moderate users of ICTs; and Advanced users of ICTs. The categorization was based on number of ICTs used by a firm. The types of ICTs used in the study are: CAD/CAM, CAE, CNC, FMS, MIS, email, Internet, and Web enabled technologies. Non-adopters were firms that were not using any of the above technologies. Moderate ICT using firms were those that were using maximum four of the above technologies whereas advanced adopters were those that were using more than four technologies. This type of classification is quite prevalent in ICTs because of hierarchical nature of technologies. For instance firms using web-enabled technologies are expected to use email, MIS, and Internet technologies.

6.1.1 Age of the sample firms

It can be seen in Table 6.17 that only 6 (9 per cent) of the 67 sample firms were established before 1970 while the majority – 58.2 per cent were established between 1981 and 2000. During the immediate post-civil war period in the early 1970s, the country enjoyed reasonable level of industrial and commercial development, thus, we note that about 27 per cent of all the sample firms were established between 1971 and 1980. With particular reference to ICT adopters and non-adopters, 10 (83.33 per cent) out of the 12 non-adopters were established before 1991 compared to 20 (66.67 per cent) out of the 30 moderate adopters and 16 (64 per cent) out of the 25 advanced adopters. We can infer from Table 6.17 that newer firms show higher level of ICT adoption.

6.1.2 Human resources profile of the sample firms

From the summary of the human resources profile of the firms by sectors presented in Tables 6.18 and 6.19, it is obvious that there is appreciable divergence in the human resource endowments. Based on previous studies on African SMEs by Liedholm (1992) and Oyeyinka-Oyelaran (1997b), firms

Table 6.17 Year of establishment of sample firms

Year of establishment of sample firms by level of ICT adoption	Level of adoption							
	Non-adopters		Moderate adopters		Advanced adopters		Total	
	No.	percent	No.	percent	No.	percent	No.	percent
Before 1970	1	1.5	2	3.0	3	4.5	6	9.0
1971–1980	6	9.0	7	10.4	5	7.5	18	26.9
1981–1990	3	4.5	11	16.4	8	11.9	22	32.8
1991–2000	2	3.0	7	10.4	8	11.9	17	25.4
After 2001			3	4.5	1	1.5	4	6.0
Total	12	17.9	30	44.8	25	37.3	67	100.0

Note: percent → is percentage of total no. of firms.

Table 6.18 Distribution of sample firms by size of employment

Size of employment	Level of adoption							
	Non-adopters		Moderate adopters		Advanced adopters		Total	
	No.	percent	No.	percent	No.	percent	No.	percent
1–9	4	6.0	5	7.5	3	4.5	12	17.9
10–40	3	4.5	18	26.9	13	19.4	34	50.7
More than 40	5	7.5	7	10.4	9	13.4	21	31.3
Total	12	17.9	30	44.8	25	37.3	67	100.0

Note: percent → is percentage of total no. of firms.

employing less than 10 persons are generally regarded as micro-enterprises. Firms employing between 10 and 49 persons are classified as small-scale, 50 to 199 as medium-scale and from 200 and above are regarded as large-scale (Adeoti, 2002). This classification is relevant to this study in which there are 12 micro-enterprises out of which 4 are non-ICT adopters, 5 (7.5 per cent) moderate adopters and 3 (4.5 per cent) advanced users. Almost 51 per cent of the sample firms employed between 10–40 persons while 31.3 per cent employed more than 40 persons.

Overall, 3,195 persons were employed by the 67 sample firms distributed as follows: Non-adopters 647 (20.25 per cent), moderate adopters 903 (28.26 per cent) and advanced adopters 1,645 (51.49 per cent). About

Table 6.19   Educational level of workforce

Academic qualification	Level of adoption							
	Non-adopters		Moderate adopters		Advanced adopters		Total	
	No.	%	No.	%	No.	%	No.	%
Engineering graduate	132	35.29	67	17.91	175	46.79	374 (11.71)	100.0
Post-graduate	120	34.99	91	26.53	132	38.43	343 (10.74)	100.0
Diploma	76	12.08	166	26.39	387	61.53	629 (19.69)	100.0
Under graduate	16	4.19	208	54.45	158	41.36	382 (11.96)	100.0
IT training	34	11.68	123	42.27	134	46.05	291 (9.11)	100.0
Others	269	22.87	248	21.09	659	56.04	1176 (36.81)	100.0
Total employees	647	20.25	903	28.26	1,645	51.49	3,195	100.0

Note: percent → is row percentage; Figures in parentheses are column percentage.

9 per cent of the entire work force received one form of ICT training or the other. Expectedly, the number of such trained workers increases with the level of adoption.

In terms of academic qualifications and total employment diploma holders predominate comprising 19.69 per cent, followed by under-graduates (11.96 per cent), engineering graduates (11.71 per cent) and post-graduate degree holders (10.74 per cent). It is perhaps worth noting that the share of IT trained staff in advanced adopters is higher than the other two groups indicating that ICT adoption actually facilitates employment creation rather than the morbid fear of job loss being ascribed to ICT by some segments of the public and labour unions.

6.2   International orientation of the sample firms

6.2.1   Existence of foreign collaboration

About half of the 67 sample firms had one form of collaboration or the other while the other half went about their business without any foreign partnership. Among the non-adopters one is engaged in collaboration with foreign partners, whereas 78 per cent and 50 per cent of the advanced and moderate adopters respectively have foreign partners (Table 6.20). Again, the dominant position of the advanced adopters supports existing literature that ICT adoption promotes outward-ness and inter-firm networks. Foreign

Table 6.20   Foreign collaboration

Foreign collaboration	Level of adoption							
	Non-adopters		Moderate adopters		Advanced adopters		Total	
	No.	percent	No.	percent	No.	percent	No.	percent
Yes	1	8.3	15	50.0	17	78.0	33	49.3
No	5	41.7	3	10.0			8	11.9
No response	6	50.0	12	40.0	8	32.0	26	38.8
Total	12	100.0	30	100.0	25	100.0	67	100.0

collaboration is crucial for the purposes of acquiring production lines, plant layout design, purchase of raw materials, training on the machines and equipment installation and test running as well as export activities among others.

Advanced adopters have an edge over the other two groups because they have higher proportion of collaborators with foreign partners. Out of the 33 enterprises having foreign patterns, advanced adopters account for 51.51 per cent followed by moderate adopters and non-adopters with 45.45 and 3.03 per cent respectively. However, some of the micro-enterprises especially those engaged in metal fabrication, ethnic clothing (textiles), wood carving and packaging of cocoa and chocolate drinks could thrive albeit at a slow pace without foreign collaboration and ICT adoption.

6.2.2   Mode of technology transfer and payment

Foreign partners provide technology either as a complete package or in parts as desired by the local partner. The technology could be in form of any or all of the following: capital goods, industrial property right such as patents, trade marks, and brand names; franchise; managerial training and organizational skills etc.

Firms with access to appropriate technology transfer and related matters through foreign collaboration stand a better chance of technological learning, adoption of new technologies and innovation. The data presented in Table 6.21 suggest that the two major modes of technology transfer imbibed were importation of machinery (26.87 per cent) and importation of technical design and drawings (26.87 per cent).

Payments for services rendered by foreign partners can take several forms: lump sum fees, running royalties, share of sales profits (net/gross), issue of equity representing the capitalization of technology, payment on 'as used' basis or 'services rendered' basis etc. The ultimate payment, however, depends on the nature and value of the technology supplied as

Table 6.21   Mode of technology transfer and payment

Mode of technology transfer and payment mode	Level of Adoption							
	Non-adopters		Moderate adopters		Advanced adopters		Total	
	No.	percent	No.	percent	No.	percent	No.	percent
Import of machinery	1	1.49	8	11.94	9	13.43	18	26.87
Import of design and drawing	5	7.46	5	7.46	8	11.94	18	26.87
Training	2	2.98	1	1.49	4	5.97	7	10.45
Lump sum	3	4.48	4	5.97	3	4.48	10	14.92
Technical fee	2	2.98	2	2.98	2	2.98	6	8.96

Note: percent → is percentage of total no. of firms.

well as services rendered and on the relative bargaining power of the two parties involved. In this study, the sample SMEs were willing to pay to their foreign technical partners in the form of technical fees and lump sum in that order.

6.2.3   *Benefits of foreign collaboration*

Partnerships between local and foreign firms (i.e. inter-firm linkages) range from subcontracting, market linkages with customers and suppliers, informal and formal collaboration (joint ventures, franchise), membership of professional and trade associations, and movement of skilled personnel from one firm to another. However, the type of linkages that predominate is as a result of the firm-level capabilities and the economic environment among others. The East and South East Asian economies have been credited to have benefit from firm interactions with external agents leading to important technological know-how and learning. Subcontracting was singled out as a key pathway for technological transfer in the garment and textile industries (Mytelka and Tesfachew, 1998). However, this process is yet to be firmly rooted in African industry (Oyeyinka, 2005).

While the quantum and quality of benefits enjoyed by Nigerian SMEs are not of the same magnitude with those of their Asian counterparts, yet, most of the firms expressed the view that certain benefits accrued to them as a result of their collaboration with foreign firms. From the data summarized in Table 6.22 we note that the most important benefit accruing to the firms is access to foreign technology (50.00 per cent), followed by goodwill

Table 6.22   Benefits of technological collaboration

Benefits of highest importance	Level of adoption							
	Non-adopters		Moderate adopters		Advanced adopters		Total	
	No.	%	No.	%	No.	%	No.	%
Foreign technology	4	66.67	6	31.58	12	57.14	22	50.00
Goodwill			3	17.65	6	31.57	9	21.43
Marketing			6	33.33	3	16.67	9	21.43
International linkages			5	27.78	4	21.05	9	20.93
Brand name			2	11.76	6	31.58	8	19.05

Note: percent → is row percentage; Figures in parentheses are column percentage.

and marketing (21.43 per cent), international linkages (20.93 per cent) and brand name (19.05 per cent). In addition to these factors, foreign partners also facilitate the import and export activities of some of the firms (Table 6.23). All these benefits are quite germane given the nature of business enterprises and the criticality of domestic and international competitiveness. These findings are generally similar to those of Oyeyinka (1997a), Adeoti (2004) and Bamkole (2005) on Nigeria.

6.2.4   *Role of foreign partners in export and import activities*

Many business enterprises are involved in export and import activities for the purposes of nurturing and sustaining their businesses and competitiveness. Nigerian SMEs for example, depend on some raw materials, machines, equipment and technical support all sourced from foreign countries especially the EU, USA and Asia. On the other hand some local firms export raw materials, such as cocoa beans and timber, intermediate products e.g. cocoa cake and finished products like apparels, plastic products, biscuits, computer parts and software.

This study has confirmed that some firms are engaged in both export and import activities. It is expected that affiliation with a foreign partner would facilitate the processes of export and import especially in the areas of market access, reliable sources of relevant equipment and machines, foreign currency, brand name information, documentation and cargo handling among others. It is instructive to note that the entire data on export and import flows are not captured in the national accounts due to smuggling and undocumented trading activities in the Central African and West African sub-regions. The ECOWAS Trade Liberalization Treaty permits free movement of goods in the latter sub-region.

Within this context the summarized data shown in Table 6.23 confirm that some firms have actually benefited from collaborative arrangements

Table 6.23   Role of foreign partners in export and import activities

Role of foreign partners in export and import	Non-adopters		Moderate adopters		Advanced adopters		Total	
	No.	percent	No.	percent	No.	percent	No.	percent
No response	5		18		12		35	
Not important	1	14.29	3	25.00	5	38.46	9	21.43
Weak	4	57.14	2	16.67			6	14.29
Fair	1	14.29	2	16.67	1	7.69	4	9.52
Very important			2	16.67	2	15.38	4	9.52
Extremely important	1	14.29	3	25.00	5	38.46	9	21.43
Total	12		30		25		67	

Note: percent → is percentage of firms to total firms responded in reach category.

with foreign firms. While 13 firms (30.95 per cent) rated the role of the foreign partners as important/very important, about ten (23.81 per cent) rated their role as weak or fairly important while nine (21.43 per cent) firms did not consider any role being played by foreign collaborator. The remaining proportions were undecided on this issue. Again, more advanced adopters rated the rate of foreign partners as being important/very important than both moderate adopters and non-adopters. This finding lends credence to the fact that successful firms under globalization need to develop and sustain domestic and international networks and partnerships. We are, however, constrained to point out that some of those firms without foreign partners might have rated their role as being important/very important because they are aware of the accruable benefits which they actually favour and desire to achieve.

6.2.5   Role of ICT in foreign collaboration

In view of the fact that ICT is very crucial in attaining efficiency, higher productivity, stock control networking and improved sales among others, it was deemed desirable to ascertain the extent to which ICT influenced collaboration between Nigerian and foreign partners. The summary of findings on this issue, which is contained in Table 6.24, indicates that advanced adopters rated highly the role of ICTs in foreign collaboration while eight firms (20.51 per cent) rated ICTs as not important for the same purpose. On the other hand, 23.07 per cent of all the firms either rated ICTs as being weak or fairly important whereas 41.79 per cent of the sample firms were undecided.

Table 6.24   Role of ICTs in foreign collaboration

Role of ICTs in foreign collaboration	Level of adoption							
	Non-adopters		Moderate adopters		Advanced adopters		Total	
	No.	%	No.	%	No.	%	No.	%
No response	3		13		12		28	
Not important	3	33.33	5	29.41			8	20.51
Weak	1	11.11					1	2.56
Fair	2	22.22	3	17.65	3	23.08	8	20.51
Very important	1	11.11	3	17.65	4	30.77	8	20.51
Extremely important	2	22.22	6	35.29	6	46.15	14	35.90
Total	12		30		25		67	

Note: percent → is percentage of firms to total firms responded in reach category.

Generally, the non-adopters did not rate the role of ICTs in this regard high, just as a handful of the moderate adopters. The main inference here is that this group of firms utilized other sources like trade exhibitions, marketing/sales brochures and catalogues, trade associations, adverts and foreign embassies as source of information, accessing and maintaining contract technical partners. It also suggests that although ICT is important, other factors apart from ICT could pay crucial roles in the decision to enter into foreign collaborative arrangements. Such factors include the need for training, access to reliable technical design and drawings, export and import activities among others.

6.2.6   Changes in product mix

Changes in product mix through diversification and or acquisition of firms producing different product lines represent progress and advances in technological capability. Hence a question in the main research instrument elicited information on this issue. The response to the question presented in Table 6.25 indicate that only two firms (3.0 per cent) either by omission or commission did not make any change in their product mix over a period of 11 years. On the other hand, every four out of ten responding firms stated that they have made average changes in their product mix within the preceding 10 years. Nonetheless, the positive influence of ICT adoption on product mix changes is confirmed by the higher proportion of advanced adopters that claimed to have made very many changes in their product mix than moderate adopters and non-adopters. These changes would have been informed by such factors as changes in consumer preference, competition, pressures from foreign partners, and threats of substitutes and adoption of ICTs in the efforts to be competitive.

Table 6.25 Changes in product mix of sample firms

Product mix changes in the firms	Non-adopters		Moderate adopters		Advanced adopters		Total	
	No.	percent	No.	percent	No.	percent	No.	percent
No change	1	8.33			1	4.00	2	2.99
Average	9	75.00	13	43.33	5	20.00	27	40.30
Very much	2	16.67	17	56.67	19	76.00	38	56.71
Total	12	100.0	30	100.0	25	100.0	67	100.0

Note: percent → is column percentage.

6.2.7 Period of ICT adoption

While the period of ICT adoption spanned a time period of more than 20 years, it is worthy of note that every six out of ten adopted, did so after 1990 (Table 6.26). The low level of adoption before then can be rationalized on the ground that until the early 1990s access to telephone services and personal computers were very much restricted in the country. Subsequently, personal computers became accessible and many business enterprises gained access to telephone despite its high cost in order to sustain their business and to be more competitive.

6.2.8 Types of ICTs adopted by the sample firms

Since several ICTs exist for potential adoption and utilization for different purposes, the sample firms have therefore adopted a variety of ICTs. It can

Table 6.26 Level and duration of ICT adoption

Period when firms started using ICTs	Non-adopters		Moderate adopters		Advanced adopters		Total	
	No.	percent	No.	percent	No.	percent	No.	percent
Before 1980			4	13.33	4	16.00	8	14.54
1980–1990			5	16.67	2	8.00	7	12.73
1991–2000			13	43.33	16	64.00	29	52.73
After 2000			8	26.67	3	12.00	11	20.00
Total			30	100.0	25	100.0	55	100.0

Note: percent → is column percentage.

Table 6.27 ICTs and their level of adoption

Major types of ICTs	Non-adopters		Moderate adopters		Advanced adopters		Total	
	No.	percent	No.	percent	No.	percent	No.	percent
CAD/CAM			2	6.67	16	64.00	18	27.87
CAE			1	3.33	9	36.00	10	14.93
Web enable technologies			1	3.33	11	44.00	12	17.91
FMS			13	43.33	20	80.00	33	49.25
CNC			9	30.00	22	88.00	31	46.27
E-mail			17	56.67	21	84.00	38	56.72
MIS			9	30.00	24	96.00	33	49.25
Internet			9	30.00	21	84.00	30	44.78
Total	12	100.0	30	100.0	25	100.00	67	100.0

Note: percent → is percentage column percentage.

be seen from Table 6.27 that among moderate adopters email (56.67 per cent) was considered most preferred technology for communication followed by FMS (43.33 per cent), CNC, MIS, and the Internet (30 per cent). Usage of CAD/CAM, CAE, and web enabled technology was very low among moderate adopters. While users of MIS were highest (96 per cent) among advanced adopters of ICTs, users of email and the Internet were substantially high (84 per cent). Technologies meant for production processes such as FMS and CNC were heavily used by advanced adopters. Although the users of CAM/CAM and CAE, and web enabled technologies were much higher to their counter part in moderate adopters of ICTs, the percentage of users of these technologies among advanced adopters of ICTs was comparatively lower than other type of technologies. The adoption of these three ICTs separated the advanced from moderate adopters. All these technologies impact positively on business enterprises. However, certain ICTs have been adopted by some firms for specific types of business activities. These findings are similar to those of Adeoti (2003), Kajogbola (2004), and Bamkole (2005).

Given the unenviable state of the physical infrastructure and low level of ICTs penetration in the country, it is not surprising that 34 (61.2 per cent) of the 55 firms that adopted various ICTs encountered one form of obstacle or the other during implementation of the adoption decision. Some of the impediments were high initial capital outlay, inadequate in-house ICT trained personnel, poor inter-connectivity, high cost of operations and maintenance.

Table 6.28   Average score of reasons for the ICT adoption

Major reasons for the adoption of ICTs	Level of Adoption			
	Non-adopters	Moderate adopters	Advanced adopters	Total
Internal competition	2.58	3.50	4.08	3.55
External competition	1.25	2.47	3.48	2.63
Flexibility in design	1.92	3.40	4.16	3.42
Reduction in production cost	2.50	2.60	4.08	3.13
Increase in sales	2.33	3.60	4.32	3.64
Increase in exports	1.42	2.43	3.28	2.57
Efficiency in production processes	2.83	3.73	4.76	3.96
Economies of scope advantage	1.50	3.37	3.96	3.25
Information about latest market trends	2.42	4.37	4.72	4.15
Better management control	2.5	4.10	4.16	3.84

Note: Score Rating: 1 = Not Important, 2 = Weak, 3 = Fair, 4= Very Important,
5 = Extremely Important

6.2.9   *Major reasons for ICT adoption*

Many cogent reasons were cited by the sample firms for ICT adoption as shown in the Table 6.28. We note that the most important reason is the use of ICTs to obtain information about latest market trends, which is a critical strategy for remaining competitive (O'Brien, 2003). Adoption of ICTs in order to enhance efficiency in production or service and better management and control are other reasons volunteered. The responses pre-

Table 6.29   Consequences of the ICT adoption

Major consequences of ICT adoption and utilization	Level of Adoption							
	Non-adopters		Moderate adopters		Advanced adopters		Total	
	No.	percent	No.	percent	No.	percent	No.	percent
Deskilling			6	20.00	10	40.00	16	29.09
Lead time advantage			20	66.67	22	88.00	42	76.36
Flexibility			23	76.67	25	100.0	48	87.27
Reorganization of management			18	60.00	22	88.00	40	72.73
Mechanization of small parts assembly			8	26.7	12	48.0	20	36.37

Note: percent → is percentage of total no. of firms.

sented in Table 6.29 indicate that most of the adopter firms are to a large extent shining examples in their various sectors and are very sensitive to the forces of competition.

6.2.10   Consequences of ICT adoption and utilization

The data in Table 6.29 summarize information on the consequences of ICT adoption as expressed by the sample firms. Although there are several potential and probable consequences of ICT utilization, the prominent consequences ranked by the firms are as follows: flexibility in production processes, lead time advantage, and reorganization of management, and mechanization of small parts assembly. These findings are generally not different from the results of the studies conducted by Kajogbola (2004) and Adeoti (2003, 2005) both of who found that ICT adoption by sampled enterprises in Nigeria led to greater efficiency of operations.

6.2.13   Challenge of competitiveness

Successful firms are very much concerned with competition and therefore take pre-emptive actions by strategizing in order to stay on top, maintain and or increase their market share. This is the situation with several of the sample firms in this study. Thus, 27 per cent of the sample recognized 41 main local competitors while 22.4 per cent and 11.8 per cent held the view that there were 34 local-based companies and 18 local-based firms with foreign collaboration respectively that are their main competitors. But the respondent firms could only identify three overseas-based main competitors. We are of the opinion that these firms have under-rated the challenge from foreign competitors in view of the fact that the local market is flooded by imported products of diverse shapes, sizes, models and brands. Many local textile firms have closed due to stiff competition from imports. The situation became an embarrassment to the government that it was forced to ban the importation of many consumer products including food items and industrial raw materials in 2004.

The vigorous domestic rivalry is a critical factor in stimulating firms to innovate in the ways that upgrade their competitive advantage. However, Mytelka (1998) noted that competitive conditions in the domestic market are not necessarily the most important driver of competitiveness. Nonetheless, most of the firms in our sample recognized the importance of several factors of competitiveness as depicted in Table 6.30.

Supply of good quality products and services received an impressive score (4.69) from 95.5 per cent of the sample closely followed by size of production/market score (4.19) by same percentage of firms, brand name/goodwill score (4.16) by 95.5 per cent of firms, and wide/good market network score (4.14) by 94.03 per cent of firms. Only about four out of every ten respondent firms claimed low salary/wage rates to be an important/very important competitive strategy which is not unexpected given the generally low

Table 6.30   Average score of strategies for competitiveness

Strategies for competitiveness	Level of adoption											
	Non-adopters			Moderate adopters			Advanced adopters			Total		
	No.	%	AS	No	%	AS	No	%	AS	No	%	AS
Quality of the product/service	11	91.67	4.91	29	96.67	4.62	24	96.0	4.67	64	95.52	4.69
Flexibility in design	10	83.33	4.10	29	96.67	3.76	23	92.0	4.09	62	92.54	3.94
Advertisement	11	91.67	2.73	29	96.67	2.90	24	96.0	3.58	64	95.52	3.13
Research and development	11	91.67	2.18	26	86.67	2.62	23		3.48	60	89.55	2.87
Size of operation/Market share	11	91.67	4.27	29	96.67	4.14	24	96.0	4.21	64	95.52	4.19
Delivery schedule	11	91.67	4.18	28	93.33	3.89	24	96.0	4.21	63	94.03	4.06
Low overhead cost	10	83.33	2.30	29	96.67	2.97	24	96.0	3.75	63	94.03	3.16
Brand Name/Goodwill	11	91.67	3.36	29	96.67	4.28	24	96.0	4.38	64	95.52	4.16
Market network	10	83.33	4.20	29	96.67	4.00	24	96.0	4.29	63	94.03	4.14
Low wage and salary	10	83.33	2.10	28	93.33	3.21	23	92.0	3.52	61	91.04	3.15
Technology collaboration	10	83.33	2.30	28	93.33	3.21	24	96.0	3.92	62	92.54	3.34
Proximity of local supplier	10	83.33	2.40	28	93.33	3.79	24	96.0	4.25	62	92.54	3.74

Note: AS → Average Score; Percentage is column percentage.
Score Rating: 1 = Not Important, 2 = Weak, 3 = Fair, 4= Very Important, 5 = Extremely Important

salary/wage in the country – a situation very much exacerbated by high unemployment, underemployment and inflation rates.

Generally, the adopters responded better to the question posed in their desire to be much more competitive. However, one of the major policy challenges arising from this finding is the relative low rating given to foreign collaboration for the purposes of gaining access to technology (52.3 per cent), and involvement in research and development (R&D) (35.8 per cent). These ratings reflect the prevailing widespread ignorance that low technological capability and innovation in the country are major causes of poor economic performance.

6.3   Statistical analysis

Having classified sample firms by their characteristics and the level of ICT adoption (IT_LEVEL) the data were analysed in a multivariate framework. We have tried to identify the causes and the consequences of the adoption of varying degree of ICT adoption by sample firms. The factors can be broadly grouped into three categories, namely: financial; firm's absorptive capacity, and factors related to globalization. Financial factor is represented by economic capacity of firm to invest in new technologies. Sales turnover (STO) in Million Naira for the year 2002–03 has been used as a proxy of financial factor. Absorptive capacity of firms has been measured by the technical workforce (TECH_EMP) of firms while globalization factor is represented by price and non-price competition such as internal and external competition (INT_COM and EXT_COM), pressure of reducing production cost (REDU_COST), flexibility in product design (FLEX_DSN), efficient production processes (EFFI_PROD), and role of market information in the business (MARK_INFO). Dependant variable being a discrete variable, it was not appropriate to use Ordinary Least Square estimation. Dependant variable measures the intensity of ICT adoption and is a ranked variable. Hence, we have preferred to use ordered PROBIT regression model. The exact specification of the model is as follows:

IT_LEVEL= f(Financial Capacity, Technological Absorptive Capacity, Globalization Factors)

The results are presented in Table 6.31. We had to use four specifications to overcome with the problem of multicolinearity (Appendix Table 6.2). In first specification all independent variables were taken together.

It can be seen from Table 6.31 that STO and EXT_COM emerged significant at the level of 10 per cent and 5 per cent respectively. In the second specification these variables were dropped. It can be seen from the results that TECH_EMP emerged significant at 5 per cent while FLEX_DSN was significant at 10 per cent level in the third model. In the fourth model INT_COM and REDU_COST were analysed. Only REDU_COST emerged

Table 6.31 Maximum likelihood estimates of ICT adoption (ordered PROBIT model)

Equations	I	II	III	IV
Constant	−1.79	−1.327	−1.0928	−.2515
Financial Capacity				
STO	0.0265*			
	(0.0782)			
Technological Absorptive Capacity				
TECH_EMP	0.0848	0.0651**		
	(0.1587)	(0.0367)		
Globalization Indicators				
INT_COM	−0.0404	−0.1102	−0.0308	0.1356
	(0.8952)	(0.5605)	(0.8581)	(0.2574)
EXT_COM	0.425**			
	(0.0194)			
FLEX_DSN	0.0580	0.2547	0.3018*	
	(0.8237)	(0.1442)	(0.0535)	
REDU_COST	0.0477	−0.06913	0.0146	0.2387*
	(0.8584)	(0.6625)	(0.9253)	(0.0402)
EFFI_PROD	−0.107	0.1556	0.5631	
	(0.8411)	(0.6185)	(0.8332)	
MARK_INFO	0.201	0.2641	0.2324	
	(0.5430)	(0.1918)	(0.2456)	
Threshold parameter for index ($\mu 1$)	2.120	1.5859	1.5196	
N	50	62	62	63
Log likelihood	−30.888	−50.173	−53.583	−59.0442

Note: Significant * at 10%, ** at 5%, and *** at 1%.
Figures in parentheses are z-values.

significant at 5 per cent level. It can be seen from Appendix Table 6.2 that INT_COM, MARK_INFO, and EFFI_PROD were correlated with almost all other independent variables. Hence they were not significant when taken along with other variables though MARK_INFO was significant at 1 per cent while INT_COM and EFFI_COM were significant at 5 per cent level when taken alone.

Emergence of STO is neither surprising nor unexpected. The finding is very much in line with the earlier studies (Lal, 2002; Siddharthan, 1992). Larger size of operation provides financial strength to invest and experiment with new technologies. Hence firms with larger size of operation are better positioned to adopt more advanced ICTs. The study clearly captures the effect of globalization on SMEs in Nigeria. They are affected mainly by two ways: one, the domestic market is exposed to multinational firms that are equipped with better technology and enjoy their international reputation and brand name. Local firms are expected to face competition from such firms. This is captured by the fact that the external competitiveness variable emerged significant. Dumping cheap products especially from China and other Asian countries is another way by which local firms were affected. Pressure on local firms was mounted to reduce the production cost by introducing efficiency in the production processes. Hence, the advanced ICT using firms reported that they adopted ICTs to achieve greater efficiency in production. Another reason for the adoption of ICTs which has emerged significant is the flexibility in product design gains of new technologies. This finding is also in line with the existing literature (Lal, 1996). Programmable feature of ICTs provides opportunities to firms to manufacture product of flexible design. This factor can also be attributed to globalization of Nigerian economy.

The study finds evidence in favour of absorptive capacity of workers in the adoption of new technologies. Absorptive capacity is represented by the number of technical employees in a firm. Although at the end user level one does not require high skills to operate ICT tools, the employees who are technology savvy are better equipped to use new technologies more effectively than others. For instance, programming knowledge of workers is an added advantage for modular production processes. Although similar findings have been reported by earlier studies (Lal, 2004; Alcorta, 1992), the measurement of absorptive capacity is different from earlier studies. It is also found that SMEs adopted ICTs for getting more appropriate and accurate business information which included market trend in term of product specifications, information about new production technologies, and exchange of other business-related information. This aspect of firms' behaviour can to some extent be attributed to globalization. Firms need to have such information as the consumers are aware of differentiated and latest products. And if they do not find those features in local products, there is fear that consumers will go in for imported products that are easily available in the domestic market. Hence firms need to update business information on a regular basis. And this is greatly facilitated by the Internet and other web enabled technologies.

7   Selected case studies

7.1   Beauty Base Limited

The company was established in Aba and started production in late 1994. Its initial product mix comprised body and laundry soaps with less than 20 workers. In October, 2004 there were 114 employees on the payroll and the product mix has changed moderately to include assorted medicated body soaps prominent among which is Terry Medicated Soap. Among the staff are three engineers, eight post-graduates, eight diploma holders, seven ITI trained and 88 unskilled workers.

Although the enterprise is entirely owned by the indigenous promoters, it has foreign collaboration. Through this arrangement the firm acquired machinery, design and drawing of factory layout and production processes. Payments for the technology transfer are in form of lump sum and agreed technical fee. Raw materials and machinery are imported from Australia, UK and Italy as need arises. Products are exported to ECOWAS and Central Africa by the firm itself, some of its distributors and traders.

The firm adopted ICTs like CAD/CAM, CAE, FMS, CNC, email, MIS, Internet and fax. The entire accounting system, staff payroll, stock control, raw materials delivery, sales and production systems are computerized. Also, the quantities and ratio of the different chemicals and perfumes for each brand of soap and detergent are determined by relevant software. The Chief Executive Officer (CEO) who has an MBA from Harvard University, USA regularly accesses market and related information from the Internet.

Some of the important/very important factors which prompted ICT adoption are competition (internal and external), flexibility in design, reduced unit cost of production, increased sales, production efficiency, access to information about market trends and better management control. Low rated factors are increased exports and opportunity to produce other products. The CEO claims that they do not rely too much on direct exports because as of now third parties export their products.

With a sales turnover of NGN 5 million in 1999/2000, it rose steadily to NGN 15 million in 2000/2001 and NGN 24.1 million in 2003/2004. Given the harsh operating environment, these values are impressive, a feat attributed to the adoption of ICTs and business re-engineering strategies by the computer literate CEO.

The management is very much aware of several competing imported soap and detergent brands as well as three main and ten secondary local competitors. In order to remain competitive in the production of the consumer goods, the firm has given great priority to good product quality, market networking, wage reduction, technology collaboration, R&D, brand name, advertisement and market share protection.

7.2   Mouka Pipe Manufacturers

The company was established in Lagos sometime in 1976 as a division of Mouka Limited because it belongs to the Moukarin family from Lebanon. The starting product lines were steel pipes and nails. On the payroll are 195 employees among which are seven engineering graduates and ten with IT training. There is an existing technical collaboration between the firm and foreign partners which supplied machinery, design and drawings for the plant in addition to training. This relationship has benefited the local firm in other spheres such as marketing, international linkages, goodwill and

ICT adoption. According to the Managing Driector, Mr. Hasib Moukarim, foreign collaboration has been sustained with ICT adoption by the firm. Due to several internal and external factors, this firm did not commence ICTs utilization until 1996. Since then however, it had adopted the email, web enabled technologies, Internet, CNC, MIS, and wireless link with mobile radios. The firm hosts a website www.mouka.com for the purposes of promoting its products.

The management of Mouka Pipe Manufacturers decided to adopt ICTs because of the high premium placed on its capability to facilitate flexibility in design, efficiency and reduction in unit cost of production, access to current market information, better management control of operations, as well as meeting the challenge of internal competition. The technology helps to prepare reports much more efficiently and timely. The quality of products has also increased over time.

Although additional production lines have opened for steel wire and zed purline, the sales turnover which rose from NGN 7 million in 1999–2000 to NGN 8.2 million in 2000–2001 fell to NGN 6.1 million in the following financial year. While revenue from exports was not released, the firm imported raw materials worth about NGN 5.5 million between 1999 and 2002.

About 30 of the entire 195 workforce are involved in ICT activities which include computerized production, accounting and staff payroll, delivery of raw materials, stock control, MIS, marketing and advertisement among others. Initially only five employees were IT trained before they were hired, the management ensured that five other suitable staff received IT training on the job.

Much as the firm downplays external competition it nevertheless strategies with approaches like improved product quality, flexibility in design, low overhead costs, brand name/goodwill, retention of market share, advertisement, technical collaboration, prompt delivery schedule, R&D and strong market network.

7.3   Louis Cater Industries Limited

Louis Cater Industries Limited (LCIL) was established in 1989 by a local entrepreneur. The initial product mix was industrial filters after which the firm diversified first into the production of vehicle exhaust pipes and subsequently paints. In October 2004 during the study interview, LCIL had 63 staff including 12 each with engineering degrees and IT training.

The establishment of the firm was facilitated by foreign technical partnership entered into by the promoter which culminated into the supply of machinery and installation in Nnewi, Anambra State. Lump sum payment was made for the transfer of technology. The promoter of LCIL rates very high the benefit of foreign collaboration in the areas of technology transfer and goodwill.

LCIL first adopted ICTs in 2000. During the interview we observed that the production process has been computerized while the ratios and batches for mixing different raw materials for paint production are also automated.

In the first year of operation, products worth NGN 1 million were exported rising by 100 per cent in the following year. Through its technical partners the firm imported raw materials worth about NGN 2 million in 2000 increasing to NGN 3 million and NGN 4 million in 2001 and 2002 respectively. The sales turnover has been growing by about 15 per cent annually since 2000.

LCIL management asserted that ICT adoption has appreciably improved operational activities including decision making and management. Some cost savings have been made especially on labour, stock control measures, advertisement, and mechanization of small parts assembly every year since production started in spite of the existence of about 12 main competitors. The major sources of the firm's competitiveness are better product quality, retention of market share, superior product delivery schedule, strong market network and proximity to local suppliers. Others include technological collaboration, brand name and flexibility in design. Flexibility in operations, improved quality, and efficient delivery systems have been attributed to ICT adoption and performance.

8 Summary and policy implications

The widely acknowledged critical role of SMEs in economic growth and development motivated the initiation of this study by UNU-MERIT to examine the general response in Nigeria to the apparent threats to SMEs posed by the globalization of production and markets. The two main policy research questions examined in the study are as follows: What strategies have proved useful in terms of enabling Nigerian SMEs to become more competitive through the use of ICTs?, How can ICTs be used to facilitate the participation of Nigerian SMEs in national and international supply chains?

The survey questionnaires retrieved from 67 SMEs in Aba, Ibadan, Lagos and Nnewi were analysed and interpreted accordingly. The profile of the sample enterprises indicates that only six (9 per cent) of the 67 firms were established before 1970 while 58.2 per cent were established between 1981 and 2000, and only 6 per cent after 2000 while remaining 26.9 per cent came into existence during 1971 and 1980.

Only 12 of the sample firms did not adopt any of the ICTs while 30 and 25 were classified as moderate and advanced adopters respectively based on the number of ICTs adopted. Overall, 3,195 persons were employed by the 67 sample firms distributed as follows: Non-adopters 647 (20.25 per cent), moderate adopters 804 (28.26 per cent) and advanced adopters 1,645 (51.49 per cent). About 9 per cent of the entire work force has received one form of ICT training or the other, and the number of such trained workers increases with the level of adoption.

About half of the 67 sample firms have one form of collaboration or the other while roughly 12 per cent had to go about their business without any foreign partnership. Among the non-adopters one firm was engaged in collaboration with foreign partners, whereas 78 per cent and 50 per cent of the advanced and moderate ICT adopters respectively have foreign partners. The two main preferred modes of technology transfer were importation of machinery and importation of technical design and drawings. Payments for the technologies acquired were usually in the form of royalties followed by technical fees and lump sum payment. Some of the major benefits emanating from foreign partnerships include access to foreign technology, goodwill, marketing, brand name and international linkage. Some of the SMEs depended on certain raw materials, machines, equipment and technical support from foreign countries especially the EU, USA and Asia. On the other hand, some local firms export raw materials, such as cocoa beans and timber, intermediate products (e.g. cocoa cake) and finished products like apparels, plastic products, biscuits, computer parts, and software. This finding lends credence to the fact that successful firms under globalization need to develop and sustain domestic and international networks and partnerships.

In general, firms rated highly the role of ICTs in foreign collaboration (44.77 per cent). Only 13.43 per cent of all the firms either rated ICTs as being not important at all or unimportant while 41.79 per cent were indecisive. Changes in product mix through diversification and or acquisition of firm(s) producing different product lines represent progress and advances in technological capability and response to competition. Four out of every ten responding firms claimed to have made average changes in their product mix within the preceding 10 years. Nonetheless, the positive influence of ICT adoption on product mix changes is confirmed by the higher proportion of advanced adopters which claimed to have made many changes in their product mix than moderate adopters and non-adopters.

While the period of ICT adoption spanned a time period of more than 20 years, it is worthy of note that six out of every ten adopters did so after 1990. The low level of adoption before then can be rationalized on the ground that prior to the early 1990s, access to telephone services and personal computers were very much restricted in the country. Fifty-five of the sample firms adopted a variety of ICTs. Email was being used by 56.72 per cent, followed by MIS and FMS (49.25 per cent), Internet (44.78 per cent), and numerically controlled machine tools (46.27 per cent). Other types of ICT adopted but to a lesser extent are CAD/CAM, CAE, and web enabled technologies. Due to the poor state of the physical infrastructure in the country and low level of ICTs penetration, 47.27 per cent of the 55 firms that adopted various ICTs encountered one form of obstacle or the other during implementation of the ICTs. Some of the impediments cited are high initial capital outlay, inadequate in-house ICT personnel, poor interconnectivity and high cost of operations and maintenance among others. With respect to competitiveness, the firms intend to meet this challenge by improving the quality of their products, flexibility in design, protection of market share, and goodwill/brand name among others.

8.1 Policy implications

SMEs development, growth and competitiveness under rapid globalization are contingent on unhindered access to information about the local and international operating environments, and challenges. However, the absence of business development services has made it impossible for such information gathering and dissemination. There is therefore an urgent need for the establishment of business development service centres in different parts of the country within the framework of SMEDAN, and relevant private sector stakeholders such as MAN, NASSI, and NASME among others. The centres should serve to educate SMEs on benefits of appropriate technologies and technical partners and facilitate access to them. Information about market demand and supply by sector and geographical location is also very critical to the competitiveness of firms. Availability of these types of information would most likely prompt them to look beyond central Africa and ECOWAS as potential export markets for their goods and services. In this regard, SMEDAN and the business development service centres, organized private sector associations and trade sections of Nigerian embassies have appropriate roles to play.

One of the major factors inhibiting ICT diffusion and intensive utilization is poor physical infrastructure such as adequate and uninterrupted electricity supply, communication connectivity infrastructure. The resultant effect is high cost of ICTs acquisition and maintenance. The government should urgently spur existing public-private partnerships (the private telephone operators) with a view to improving the state of these types of infrastructure.

The Small and Medium Enterprises Equity Investment Scheme as at November, 2004 has a loan-able deposit of about NGN 29 billion. Yet, many SMEs in the country are handicapped by inadequate finance. The Central Bank of Nigeria that has custody of the funds should in partnership with SMEDAN and relevant private sector stakeholders, organize seminars with a view to sensitizing SMEs on the benefits and modalities of accessing SMEEIS fund for sustainable growth and ICTs acquisition and use among others.

Since ICT is a major driver of the globalization process, for Nigerian SMEs to be competitive, they must have reasonable level of competence in ICT applications. Consequently, trade and business associations with the active support of relevant government agencies should organize regular training workshops on the various facets of e-business. Furthermore, the efficiency of business decisions is known to be contingent on the ability of managers to assess potential opportunities and risks, and to identify and select most appropriate actions to pursue business objectives subject to external and internal constraints. There should therefore be deliberate entrepreneur skills improvement programmes for SMEs.

Under the current globalization, it is no longer only individual firms that are engaged in competition rather, what we have are industrial clusters, groups of firms organized in networks, whose dynamic development and competitiveness depend on the spatial location and proximity (real and virtual) with R&D facilities, technology formation and dissemination institutions, knowledge centres, financial and export information institutions. Consequently, the government should provide an enabling environment for SMEs to operate in efficient ways and access to relevant research and development results and market information so as to make them competitive.

The global ICT revolution offers Nigeria the unique opportunity of actively participating in a globalized economy. The biggest beneficiaries are countries that are proactive and have identified the strategic relevance of ICTs in the rapid transformation of national economic development. A major policy challenge therefore is how the country could re-engineer and re-position its formal and informal educational and training systems in order to rapidly raise ICT literacy level. With particular reference to SMEs, ICT training sessions could be organized in or around their clusters. The academic curricular in the universities and polytechnics should be geared towards graduates who would take active interest in the production of computer hardware and software applications and entrepreneurship. Appropriate fiscal and monetary policies should be put in place to reduce the costs of ICT importation, production, acquisition and maintenance.

Appendix Table 6.1   Nigeria's exports[16] by products, 1999–2003 (NGN million)

Description	1999	2000	2001	2002	2003
Live animals & animal Products	968.2	338.4	27.2	62.7	152.7
Of which: Live animals	2.3	0.2	0.0	0.8	
Vegetable products	1,267.9	3,015.4	143.0	6,978.3	200.9
Fats and oils	226.6	259.9	102.8	106.1	201.6
Food industry products	2,287.2	299.6	119.7	13,337.5	393.7
Of which: Tobacco products	0.4	0	0.5	0.9	
Minerals products	1,543,512.9	2,740,488.7	2,000,454.4	2,105,872.4	3,044,175.6
Chemical & allied products	142.1	96.9	371.5	3,193.7	1,110.7
Plastics, ethers, esters of cellulose, rubber etc	612.9	201.7	277.6	2,823.6	2,725.1
Hides, leather and fur	557.8	176.9	189.6	34.2	773.6
Wood, charcoal & wood products	1,610.5	43.1	1.2	1,012.6	11.0
Paper-making material and articles	24.5	24.1	118.1	60.3	100.4
Textiles and textile articles	1,259.3	1,100.8	3,293.8	16,497.3	1,338.2
Footwear, headgear, umbrellas, feathers, hair	57.7	129.4	14.6	47.7	407.2
Stone, plaster, cement, asbestos, Mica products	302.9	486.7	51.5	37.5	81.9
Natural pearls, gemstone and other precious metals	3.5	–	–	–	–
Base metals and articles of base metal	207.3	901.8	199.8	1,394.7	1,200.3
Machinery and appliances (other than electrical)	18.1	1,801.1	198.6	2,491.3	8,938.5
Transport equipment	4,358.7	1,985.7	1,553.1	69,953	46,577.9
Instruments and apparatus (photo, clocks, etc)	90.4	24.9	0.7	90.6	659.5
Miscellaneous manufactured articles	–	23.5	7.6	15,388.2	239.2
Arms and ammunitions	–	–	–	–	–
Works of art, collectors pieces and antiques	–	–	1.1	–	–
Other goods and products	28.1	132.8	1.2	4.7	0.3
Total	1,557,539.3	2,751,531.6	2,007,127.6	2,239,388.1	3,109,288.3

Source: Federal Office of Statistics (2004).

16   Export is on the basis of Free-On-Board (FOB). This is the cost of goods to the purchaser abroad, including packaging and all other relevant charges made up to the time of their delivery on board the exporting vessel. For crude oil lifting, the realized price of the official selling price shown in the oil returns of the lifting companies is applied to the net quantify figures.

Appendix Table 6.2   Correlation coefficients

	STO	ICT_LEVEL	TECH_EMP	INT_COM	EXT_COM	FLEX_DSN	REDU_COST	EFFI_PROD
STO	1							
ICT_LEVEL	0.395**	1						
TECH_EMP	0.487**	0.328**	1					
INT_COM	.191	0.268*	.259*	1				
EXT_COM	.054	0.443**	.218	.265*	1			
FLEX_DSN	0.287*	0.474**	.363**	.586**	.469**	1		
REDU_COST	.076	0.390**	.146	.435**	.423**	.603**	1	
EFFI_PROD	.101	0.407**	.233	.646**	.524**	.700**	.716**	1
MARK_INFO	.199	0.422**	.180	.427**	.182	.446**	.501**	.534**

Note: ** → 1% and * → 5% level of significance.

References

Adeboye, T. (1995) Technological Capabilities in Small and Medium Industry Clusters: Current Knowledge and Ignorance. African Technology Policy Studies Network, Nairobi.

Adeoti, J.O. (2002) Technology and the Environment in Sub-Saharan Africa: Emerging Trends in the Nigerian Manufacturing Industry, Ashgate Publishing Ltd., Aldershot.

Adeoti, J.O. (2003) 'Information Technology as Stimulus of Manufacturing Competitiveness in Nigeria'. Interim Research Report to the African Technology Policy Studies Network, Nairobi.

Adeoti, J.O. (2004) 'Information Technology Investment in the Nigerian Manufacturing Industry: The Progress So Far'. Paper Presented at the 45th Annual Conference of the Nigerian Economic Society.

Adeoti, J.O. (2005) 'The Adoption of GSM Lines by Small-Scale Enterprises in Ibadan Metropolis, Nigeria'. NISER Discussion Paper, NISER, Ibadan.

Adeoye, B.W. (2002) 'Promoting Small Scale Enterprises for Economic Development in Nigeria'. *NISEREEL*, No. 3, Nigerian Institute of Social and Economic Research, Ibadan.

Aguilar, J. and Cordero, J.T.S. (1998). Estrategias Tecnológicasy Desempeño Exportador de la Pequeña y Mediana empresa: El Caso de Costa Rica. En Serie Divulgación Económica, 31.

Ajayi, I. (2004) 'Issues of Globalization in Africa: The Opportunities and the Challenges', *Ibadan Journal of the Social Sciences*, Vol. 2(1): 23–42.

Akerele, W.O. (2003) *Learning and Technological Knowledge Acquisition in Nigeria's Small and Medium Enterprises: The Case Study of Manufacturing Enterprises in Oyo State.* NISER Monograph Series No. 7, NISER, Ibadan.

Akinbinu, A. (2003) *Industrial Reorganization for Innovation: Current Knowledge on Small and Medium Enterprises Clusters in Western Nigeria.* Monograph Series No. 8, NISER, Ibadan.

Akinboyo, L.O. (2004) 'Globalization, Information Technology and the Nigerian Financial System'. Paper Presented at the 2003 Annual Conference of the Nigerian Economic Society (NES), Ibadan.

Alanen, L. (1996) *The Impact of Environmental Cost International on Sectoral Competitiveness: A New Conceptual Framework UNCTAD Discussion Paper No. 119, UNCTAD, Geneva.*

Alcorta, L. (1992) *The Impact of New Technologies on Scale in Manufacturing Industry.* UNU/INTECH Working paper No. 5, The United Nations University, Institute for New Technologies, Maastricht, June.

Alos, J. (2001) 'Key Drivers of Small and Medium Enterprise (SME) Development in Nigeria'. Paper presented to the SME working Group meeting at the Nigeria Economic Summit, Abuja.

Ali, A. and Sweirez, P.M. (1991) 'Firm Size and Export Behaviour: Lessons from the Midwest', *Journal of Small Business Management*, Vol. 6, pp. 71–8.

Aragba-Ekpore, S. (2005) 'Protesters Lay Siege to MTN Offices, Consumer Council Gives Deadline', *The Guardian*, 11th May, 2005, p. 80.

Bamkole, P. (2005) 'Business Development Service Provision in Nigeria: Challenges and Prospects'. In: Eric C. Eboh (ed.). *Promoting Nigeria's Non-Oil Private Sector:*

Evidence and Recommendations. African Institute for Applied Economics for the Better Business Initiative, Enugu.

Barnett, A. (1995) 'Technology and Small Scale Production', *Small Enterprise Development*, 6(4): 14–22.

Bayes, A. and von Braun, J. (1999) 'Village pay phones and poverty reduction: Insights from a Grameen Bank initiative in Bangladesh', Discussion Paper 8 (Centre for Development Research [ZEF], Bonn).

Beaver, G. (2002) Small Business, Entrepreneurship and Enterprise Development, Pearson Education Limited, UK.

Bedi, Arjun, S. (1999) 'The Role of Information and Communication Technologies in Economic Development: A Partial Survey', Discussion Paper 7 (Centre for Development Research [ZEF], Bonn).

Bell, M. and Pavit, K. (1993) 'Accumulating Technological Capability in Developing Countries', *Industrial and Corporate Change*, Vol. 2(2), pp. 35–44.

Bessant, J. (1999) 'Getting the Tail to Wag: Enabling Innovation in Small and Medium – sized Enterprises'. In: A. Szirmai, J. Halman and B. Verspagen (eds), Innovation in Theory and Practice, Papers of the ECIS Opening Conference, Technische Universiteit, Eindhoven, The Netherlands.

Boeke, J.H. (1971) 'Batas-batas Dari Masyarakat Pedesaan, Bharata, Djakarta'.

Brynjolfsson, E. and Hitt, L. (1996) 'Paradox Lost? Firm-level Evidence on the Returns to Information System Spending', *Management Science*, 42(4): 541–58.

Cane, A. (1992) 'Information Technology and Competitive Advantage: Lessons from the Developed Countries', *World Development*, 20 (12).

Castillo, G. and Chaves, L. (2003) Pymes: Una oportunidad de Desarrollo para Costa Rica. FUNDES, San José, Costa Rica.

CBN (Central Bank of Nigeria) (2003) *Annual Report and Statement of Accounts*. CBN, Abuja.

CBN (Central Bank of Nigeria) (2004) *Annual Report and Statement of Accounts*. CBN, Abuja.

Cohen, B. (1995) Empirical Studies of Innovative Activity. In: Paul Stoneman (ed.), Handbook of the Economics of Innovation and Technological Change (Blackwell, Oxford) pp. 182–264.

Court, J. and Yanagihara, T. (1998) 'Asia and Africa into the Global Economy: Background and Introduction'. Paper presented at the UNU-AERC Conference on 'Asia and Africa: in the Global Economy', 3–4 August, Tokyo, Japan.

Department of Statistics (2000) Census 2000 Report, Government of Malaysia, Kuala Lumpur.

Doms, M., Dunne, T. and Troske, K.R. (1997) 'Workers, wages and technology', *Quarterly Journal of Economics*, CXII (February): 253–90.

Dosi, G., Freeman, C., Nelson, R., Silverberg, G. and Soete, L. (eds) (1988) *Technological Change and Economic Theory*, London: Pinter Publishers Limited.

Earl, M.J. (1989) Management Strategies for Information Technology, Business Information Technology Series (Prentice-Hall, N.J.).

Ernst, D., Mytelka, L. and Ganiatsos, T. (1994) 'Technological Capabilities: A Conceptual Framework'. Chapter 1 of A *Research Report on Technical Development in South and East Asia*, UNCTAD, Geneva.

Evans, P.B. and Wurster, T.S. (1997) Strategies and the new economies of information, in: C.W. Stern and G. Stalk Jr. (eds), Perspectives on Strategies from The Boston Consulting Group (John Wiley & Sons, New York).

Federal Office of Statistics (2004) Basic Statistical Data on Nigeria: 2004, Abuja.

Federal Ministry of Science and Technology (2001) *Nigerian National Policy for Information Technology*. Federal Ministry of Science and Technology, Abuja.

Fisk, E.K. (1962) The Economy of The Handloom Industry of The East Coast of Malaysia, The Malayan Branch of the Royal Asiatic Society, Singapore.

FRN (Federal Republic of Nigeria) (2004) Basic Statistical Data on Nigeria: 2004 Federal office of Statistics, Abuja.

Fryer, D.W. (1978) The Development of Cottage & Small Scale Industries In Malaya & In South East Asia.

Gibbon, P. (2002) Present Day Capitalism, the International trade regime & Africa, *Review of Political Economy*, 91: 95–112.

Gómez, M. (2002) Competitividad de las Pymes: ¿Cómo pueden las pequeñas y medianas empresas de Costa Rica competir en el mercado local e internacional?. En Economía y Sociedad 20 pp. 127–43.

Gray, H. and Prahalad, C.K. (1994) Competing for the Future (Harvard Business School Press, Boston).

IMF (International Monetary Fund) (1997) World Economic Outlook: Globalization Opportunities and Challenges. IMF, Washington, D.C.

IT and Telecom Digest (2005) 'The Next.ng Challenge', *IT & Telecom Digest*, April p. 5.

IT and Telecom Digest (2005) 'Investment in ICT Exceeds US $4 billion', *IT & Telecom Digest*, April, p. 55.

Julien, P.A., Joyal, A. and Deshales, L. (1993) 'SMEs and International Competition: Free Trade or Globalization', *Journal of Small Business Management*, Vol. 32, pp. 52–64.

Julien, P.A. and Carriere, J.B. (1993) 'Efficacite des PME et Nouvelles Technologies', *Revue d'economie Industrielle*, Vol. 67, pp. 119–34.

Julien, P.-A. (2001) 'Globalization of Markets and Behaviour of Manufacturing SMEs'. www.statean.co/ca/english/freepub/61-532-XIE/05/Julien.

Kajogbola, D.O. (2004) *The Impact of Information Technology on the Nigerian Economy: A Study of Manufacturing and Services Sectors in the South Western and South Eastern Zones of Nigeria*. African Technology Policy Studies Working Paper Series No. 39, ATPS, Nairobi.

Kiiski, S. and Pohjola, M. (2002) 'Cross-country Diffusion of the Internet', *Information Economics and Policy*, 14(2), 276–97.

Kim, L. (1997) *Imitation to Innovation: The Dynamics of Korea's Technological Learning*, Harvard University Press, Boston.

Kirkbride, P. (2001) (ed.) *Globalization: The External Pressures*. John Wiley, Chicester, West Sussex.

Kose, A., Eswar, M., Prasad, S. and Terrones, M.E. (2004) 'Taking the Plunge Without Getting Hurt', *Finance and Development*, December, pp. 44–7.

Lal, K. (1996) 'Information Technology, International Orientation and Performance: A case study of electrical & electronic goods manufacturing firms in India', *Information Economics and Policy*, 8(3): 269–80.

Lal, K. (2001) 'Institutional Environment and Growth of ICT in India', *The Information Society*, 17: 105–17.

Lal, K. (2002) 'E-business and Manufacturing Sector: A Study of Small and Medium-sized Enterprises in India', *Research Policy*, 31(7): 1199–211.

Lal, K. (2004) 'E-business and Export Behavior: Evidence from Indian Firms', *World Development*, 32(3): 505–17.

Lal, K. (2006) 'Institutional Environment and Development of Information and Communication Technology in India', in S.D. Tendulkar *et al.* (eds)

India: Industrialization in a Reforming Economy, Academic Foundation, Delhi, pp. 225–52.

Lall, S. (1983) Determinants of R&D in an LDC, Economic Letters 13, 379–83.

Lee, H.G. and Clark, T. (1997) 'Market process reengineering through electronic market systems: opportunities and challenges', *Journal of Management Information Systems*, 13(3), 13–30.

Lee, K.J. (2003) Credit Facilities for Small and Medium Industries: Options and Opportunities, Utusan, Malaysia.

Lee, K.J. (2003) Credit Facilities for Small and Medium Industries: Options and Opportunities, Utusan, Malaysia.

Lichtenberg, F. (1995) 'The Output Contribution of Computer Equipment and Personnel: A Firm Level Analysis', *Economics of Innovation and New Technology*, 3 (3–4), 201–17.

Liedholm, C. (1992) 'Small Scale Industry in Africa: Dynamic Issues and the Role of Policy'. In: F. Stewart; S. Lall and S. Wangwe (eds), *Alternative Development Strategies in Sub-Saharan Africa.* Macmillan, London.

Little, I.M.D., Mazumdar, D. and Page Jr., J.M. (1989) *Small Manufacturing Enterprises: A comparative study of India and other countries*, Washington: IBRD/World Bank.

Lucas Jr., R.E. (1988) 'On the Mechanics of Economic Development', *Journal of Monetary Economics*, 12: 3–42.

Lundvall, K. and Battese, G.E. (2000) 'Firm size, Age and Efficiency: Evidence From Kenyan Manufacturing Firms', *The Journal of Development Studies*, Vol. 36, No. 3, pp. 146–63.

Malone, T., Yates, J. and Benjamin, R. (1987) Electronic Markets and Hierarchies, Communication of the ACM 30(6), 484–97.

Mankiw, N.G., Romer, D. and Weil, D. (1992) 'A contribution to the empirics of economic growth', *Quarterly Journal of Economics*, 107: 407–37.

Mansell, R. and Wehn, U. (eds) (1998) Knowledge societies: Information technology for Sustainable Development (Oxford University Press, Oxford).

Marcelle, G. (2004) *Technological Learning: A Strategic Imperative for Firms in the Developing World*, Edward Elgar, Cheltenham.

Marcotte, C. and Julien, P.A. (1994) 'Partage d'Information et Performance des Coentre-prizes Implantees Par Les PME Quebocoises Dan Les Pays en Development', *Revue Internationale* PME. Vol. 8, p. 8175 et Vol. 2, p. 2002, cited by Julien (2001), Ibid.

McFarlane, C. (1997) The 1996 Micro and Small Enterprise (MSE) survey of Jamaica (Government of Jamaica/Government of Netherlands Micro-Enterprise Project (GOJ/GON/MEP) and USAID, Government of Jamaica.

Mehta, D. (2000) Internet and e-commerce: The global scenario, In D. Mehta (ed.), *E-commerce and e-business: Background and reference resource*, pp. 15–19, New Delhi: NAASCOM.

MEIC (Ministry of Economy, Industry and Commerce) (1998) Manufacturing sector database 1998, Ministry of Economy, Industry and Commerce San José, Costa Rica.

MEIC (2005) Manufacturing sector database 2005, Ministry of Economy, Industry and Commerce, San José, Costa Rica.

MITI (Ministry of International Trade and Industry) (2003) Annual report, Government of Malaysia, Kuala Lumpur.

Mitra, A. (1999) 'Agglomeration Economies As Manifested in Technical Efficiency of Firms', Journal of Urban Economics, 45.

Mitra, A., Varoudakis, A. and Veganzones, M.A. (2002) 'Productivity and Technical Efficiency in Indian States' Manufacturing: The Role of Infrastructure', *Economic Development and Cultural Change*, 50.

Moreno, L. (1997) 'The determinants of Spanish industrial exports to the European Union', *Applied Economics*, 29: 723–32.

Muñoz, J.J. (2002) Política Industrial y Ajuste Estructural en Costa Rica. PhD Dissertation, Tilburg University.

Mytelka, L.K. (1998) *Competition, Innovation and Competitiveness: A Framework for Analysis.* Development Centre Studies, OECD, Paris.

Mytelka, L.K. and Tesfachew, T. (1998) *The Role of Policy in Promoting Enterprise Learning During Early Industrializations: Lessons for African Countries.* UNCTAD/GDS/MDPB/Misc. 7, UNCTAD, Geneva.

NASSCOM (National Association of Software and Services Companies) (2000) E-Commerce and E-Business: Background and Reference Resource (NASSCOM, New Delhi).

National Productivity Corporation (2003) SME Statistics, Government of Malaysia, Kuala Lumpur

Neilson, G., Pasternack, B. and Visco, A. (2002) 'Up the E-Organization! A Seven-Dimensional Model of the Centerless Enterprise', *Strategy and Business,* First Quarter, p. 53.

NIP (National Industrial Policy) (1996) A strategic Plan for Growth and Development, Government of Jamaica.

NIPC (Nigerian Investment Promotion Council) (2002) 'Overview of Small and Medium Scale Enterprise'. In: *Overview of the Nigerian Economy.* Nigerian Investment Promotion Council, Abuja.

NISER (Nigerian Institute of Social and Economic Research) (1998–2003) *NISER Survey of Business Conditions, Experience and Expectations in the Manufacturing Sector,* NISER, Ibadan.

O'Brien, J.A. (2003) *Introduction to Information Systems: Essentials for the e-Business Enterprise,* Boston: McGraw-Hill Irwin.

OECD (Organization for Economic Corporation and Development) (1997) The World in 2020. Towards a New Global Age, OECD, Paris.

Olayiwola, K. and Ogundiran, O. (2004) 'Positioning Nigeria in the Knowledge Society in a Globalized Era'. Proceedings of the 2003 Annual Conference of the Nigerian Economic Society (NES), Ibadan.

Olokesusi, F. (1987) 'Water Supply: Possible Constraints on Socio-Economic Development in Oyo State of Nigeria', *AQUA*, 5, pp. 268–73.

Osoba, A.M. (1987) *Towards the Development of Small-Scale Industries in Nigeria.* Nigerian Institute of Social and Economic Research, Ibadan.

Oyelaran-Oyeyinka, B. and Lal, K. (2005) Internet diffusion in Sub-Saharan Africa: a cross-country Analysis, *Telecommunications Policy*, 29(7): 507–27.

Oyelaran-Oyeyinka, B. and Lal, K. (2006) *SMEs and New Technologies: Learning E-Business and Development,* Palgrave-Macmillan.

Oyeyinka-Oyelaran, B. (1997a) *Nnewi: An Emergent Industrial Cluster in Nigeria.* Technopol Publishers, Ibadan.

Oyeyinka-Oyelaran, B. (1997b) 'Technological Learning in African Industry: A Study of Engineering Firms in Nigeria', *Science and Public Policy*, 24(5), pp. 309–18.

Oyeyinka-Oyelaran, B. (2004) *How can Africa Benefit from Globalization?: Global Governance of Technology and Africa's Global Exclusion,* African Technology Policy Studies Special Paper Series, No. 17, ATPS, Nairobi.

Oyeyinka-Oyelaran, B. (2005) *Networking Technical Change and Industrialization: The Case of Small and Medium Firms in Nigeria.* ATPS, Nairobi.

Pacheco, J. and Francisco, J. (1998) Política de Competitividad Industrial en Costa Rica: análisis crítico delos programas de reconversión industrial 1982–

1996. Proyecto de Graduación en Economía. Heredia, Costa Rica: Universidad Nacional.

Pavitt, K., Robson, M. and Townsend, J. (1987) 'The Size Distribution of Innovating Firms in the UK: 1945–1983', *Journal of Industrial Economics*, 35, 297–316.

Planning Institute of Jamaica (PIOJ) (2003) Economic and Social Survey Jamaica Micro and Small Enterprise, Government of Jamaica.

Pohjola, M. (ed.) (2001) Information Technology, Productivity and Economic Growth: International Evidence and Implications for Economic Development (Oxford University Press, New York).

Pohjola, M. (2001) Information Technology and Economic Growth: A Cross-Country Analysis, in: Matti Pohjola (ed.) Information Technology, Productivity and Economic Growth: International Evidence and Implications for Economic Development (Oxford University Press, New York) 242–56.

Porter, M.E. (1990) *The Competitive Advantage of Nations*, New York: Free Press.

Priestee, T.J., Llisterri, T.J. and Rivas, G. (1998) Inter-American Development Bank Group: Activities Supporting Small and Medium sized enterprises (1990–98), Jamaican documentation centre (JAMDOC), 99-04-8400.

Rada, J.F. (1982) *The impact of microelectronics and information technology: Case studies in Latin America* (UNESCO Publication, Vendome, Imprimerie Presses Universitaires de France).

Rebelo, S. (2001) The Role of Knowledge and Capital in Economic Growth, in: Matti Pohjola, (ed.) Information Technology, Productivity and Economic Growth: International Evidence and Implications for Economic Development (Oxford University Press, New York) 33–49.

Rosegger, G. (1989) The Economics of Production and Innovation: An Industrial Perspective, Pogamon Press, Oxford.

Sengenberger, W.M., Piore, M. and Pyke, F. (1990) 'The Industrial Districts and SMEs', *Institut International des Etudes Socialies*. MIT, Geneva.

Seow, L. (1989) The Role Of SMIs In National Development. AJDM Seminar, July, 1989.

Siddharthan, N.S. (1992) 'Transaction costs, technology transfer and in-house R&D: A study of the Indian private corporate sector', *Journal of economic behaviour and organizations*, 18, 265–71.

SMIDEC (Small & Medium Industries Development Corporation) (2002). SMI Development Plan. Percetakan Nasional Malaysia Bhd (PNMB).

SMI Association Malaysia (2004) SMI Business Directory – 2004, 3rd (ed.) Malaysia.

Soete (1997) Building the European Information Society for us all, Final policy report of the high-level expert group submitted to Director-General for employment, industrial relations and social affairs, European Commission (MERIT, Maastricht).

Solow, R. (1956) 'A contribution to the theory of economic growth', *Quarterly Journal of Economics*, 70: 65–94.

Stiglitz, J.E. (1989) 'Economic organization, information and development', in: J. Behrman and T.N. Srinivasan (eds), *Handbook of Development of Economics*, Vol. 1 (North-Holland, Amsterdam) pp. 93–160.

Stiglitz, J.E. (1989) 'Economic organization, information and development', in: *Handbook of Development of Economics*, Vol. 1, Hollis Chenery and T.N. Srinivasan (eds), North-Holland, Amsterdam, pp. 93–160.

Stiglitz, J. (2001) Globalization and Its Discontents. Allen Lane, New York.

Stiroh, K.J. (2001) 'Investing in Information Technology: Productivity Payoffs for US Industries', *Current Issues in Economics and Finance*, Federal Reserve Bank of New York, June.

The Guardian, 2004, 13th December, pp. 21–3.

The Punch, 2004, 9th September, p. 19.

The Punch, 2005, 3rd May, p. 58.

UNCTAD (United Nations Conference on Trade and Development) (2000) Enhancing the competitiveness of SMEs through linkages UNCTAD TD/B/COM3/em/11.2.

UNCTAD (2003) *E-Commerce and Development Report, 2003*. Internet Edition, UNCTAD, Geneva.

UNCTAD (2004a) World Investment Report, UNCTAD, Geneva.

UNCTAD (2004b) *E-Commerce and Development Report, 2004*. Internet Edition, UNCTAD, Geneva.

UNESCO/AIDCOM (2001) Socio-Cultural Research on Information, Education And Communication and Advocacy: A Practical Guide, UNESCO/Asian Institute for Development Communication (Aidcom), KL.

Verhoeven, W. (1988) 'The Export Performance of Small and Medium-sized Enterprises in the Netherlands'. *International Small Business Journal*, Vol. 6, pp. 20–33.

Wei, S.-J. (2002) 'Is Globalization Good for the Poor in China?', *Finance and Development*, September, pp. 26–9.

World Bank (1998) World Development Report (Oxford University Press, New York).

World Bank (2001) World Development Indicators: CD-ROM World Bank, Washington, D.C.

World Bank (2004) World Development Report 2004, World Bank, Washington.

Index